George M. Dawson

On the Later Physiographical Geology of the Rocky Mountain Region in Canada

With Special Reference to Changes in Elevation and the History of the Glacial Period

George M. Dawson

On the Later Physiographical Geology of the Rocky Mountain Region in Canada
With Special Reference to Changes in Elevation and the History of the Glacial Period

ISBN/EAN: 9783743418400

Manufactured in Europe, USA, Canada, Australia, Japa

Cover: Foto ©berggeist007 / pixelio.de

Manufactured and distributed by brebook publishing software (www.brebook.com)

George M. Dawson

On the Later Physiographical Geology of the Rocky Mountain Region in Canada

ON THE

LATER PHYSIOGRAPHICAL GEOLOGY OF THE ROCKY MOUNTAIN REGION IN CANADA, WITH SPECIAL REFERENCE TO CHANGES IN ELEVATION AND THE HISTORY OF THE GLACIAL PERIOD.

By GEORGE M. DAWSON, LL.D., F.G.S.

Assistant Director Geological Survey of Canada.

I.—*On the later Physiographical Geology of the Rocky Mountain Region in Canada, with special reference to Changes in Elevation and to the History of the Glacial Period ; being the Presidential Address for the year.*

By GEORGE M. DAWSON, D.Sc., A.R.S.M., F.G.S., *Assistant Director Geological Survey of Canada.*

(Read May 29, 1890.)

I.

MESOZOIC AND TERTIARY HISTORY.

Very great difficulty is found, even in the best known regions of Europe, in tracing out in detail the various phases of the physical geography of any particular geologically complicated region, from the early date at which the oldest known rocks were formed to the present time. The data for any such attempt, in the part of the North American continent here treated of, are as yet extremely fragmentary and insufficient, but as my object is merely to follow, as far as possible, the origin of its existing geographical and orographical features, it will be unnecessary to go back very far in geological time, of which the later events are already moderately well known even in the region in question. The actual features of this western country depend directly upon these later events, being much newer in origin as well as larger and bolder in outline than those of the eastern part of the continent. It is in fact chiefly on the relative recency of the orogenic disturbances of the region of the west coast, that its present bold characters depend. Such disturbances of the crust of the earth in the eastern part of the continent occurred, for the most part, in earlier periods, and their more striking results had disappeared or had become greatly reduced under the influences of natural processes of waste, even before that middle time in geological history from which I propose to trace briefly the physical history of the West. In so doing, many points are found to occur upon which much further evidence is desirable, but a formulation of the main sequence of events, in so far as known, will at least have the effect of drawing attention to such points and of stimulating further enquiry.

At the present day, the western border region of the continent is formed by a series of more or less nearly parallel mountain-systems with an average breadth in British Columbia of about 400 miles. The trend of these systems is north-west and south-east, or similar to that of the corresponding portion of the Pacific shore-line, the position of which in fact depends upon that of these orographic features. This generally mountainous zone of country is, as a whole, often referred to as the Rocky Mountains, but is more appropriately named the Cordillera belt, the Rocky Mountains proper constituting only its north-eastern marginal range. In traversing it from east to west, in the southern part of the Province of British Columbia, four distinct mountain-systems are crossed :—(1) The Rocky

Mountains proper, (2) Mountains which may be classed together as the Gold Ranges, (3) The Coast Ranges, (4) An irregular mountain-system which in its unsubmerged parts constitutes Vancouver Island and the Queen Charlotte Islands, and which may be designated the Vancouver system. Between the second and third of these mountain-systems is a region without important mountain ranges, which is referred to as the Interior Plateau of British Columbia.

To simplify our conception of the main features of this part of the Cordillera for our present purpose, we may, however, regard it broadly as being outlined on the north-east and south-west sides by the Rocky Mountains proper and by the Coast Ranges, as dominant systems. This view is justified by the remarkable constancy of these two systems and their relative importance. The intervening region may then be described as comprising the Interior Plateau together with the various ranges which have been grouped together under the name of the Gold Ranges, as well as other detached mountains and irregular mountainous tracts.

The Cordillera belt, taken in the aggregate, constitutes an elevated region, of which the broken Vancouver system of mountains forms the south-western border, based upon the edge of the continental plateau, which, at a short distance beyond it, shelves rapidly down from moderate soundings to the abyssal depths of the Pacific basin. On the other side, the north-eastern base of the Rocky Mountains is bordered by a narrow zone of foot-hills, beyond which, in easterly and north-easterly directions, the Great Plains slope gradually down to the base of the old Laurentian plateau. The elevation of these plains at the eastern base of the Rocky Mountains, in the vicinity of the forty-ninth parallel of latitude, is about 4,000 feet, about 3,000 feet only near the fifty-sixth parallel, and is even less still further north.

The average breadth of the Rocky Mountain Range proper in the southern part of British Columbia is about sixty miles, but the range decreases near the Peace River to twenty miles or less in width, and apparently loses its importance and regularity locally where cut through by the Liard River, though recovering both still further to the north-westward. Near the forty-ninth parallel, several summits occur in this range which exceed 10,000 feet in height, but northward, few attain to this height till the head-waters of the Bow River are reached. About the sources of the North Saskatchewan and Athabasca, the range appears to culminate, and Mounts Brown and Murchison occur, with reputed heights of 16,000 and 13,500 feet respectively. Near the Peace, few summits exceed 6,000 feet, so far as known.[1]

The Coast system of mountains runs without any important break or change in character from the vicinity of the mouth of the Fraser River[2] for over 900 miles in a north-westward direction. This remarkable mountain-system has an average width, nearly uniformly maintained, of about one hundred miles. The mean altitude of the higher summits is probably between 7,000 and 8,000 feet, while some points are known to exceed 9,000 feet.

The mountains of the Gold system, including the Selkirk, Purcell, Columbia, Cariboo

---

[1] I quote here and in the notice of the Coast Ranges, in part from my previous description of these features. See particularly "Mineral Wealth of British Columbia," 'Annual Report Geol. Surv. Can., 1887-88, Part R.'

[2] Where it replaces and partially overlaps the Cascade Mountains of Washington and Oregon, with which it should not be confounded.

and other ranges, lie between the Interior Plateau and the south-western base of the Rocky Mountains proper, from which they are separated by a long and straight valley only. Many peaks in these ranges are scarcely inferior in altitude to those of the Rocky Mountains themselves, and the subsidiary position accorded to them in this very general description of the country, is justified only by their more irregular character and absence of orographic continuity. Some of the rocks entering into the composition of this mountain-system are regarded as Archæan, and these rocks are more continuous along the line of their strike than the mountains themselves. It is probable indeed that this mountain axis is of greater age, as a physical feature, than either the Rocky Mountains or the Coast Ranges.

The Interior Plateau of British Columbia will be found to possess particular importance in the present enquiry. It has an average width of about one hundred miles, and I have described it in previous publications as having a mean elevation of about 3,500 feet. It does not, however, maintain the same continuity in a north-west and south-east bearing as do the Rocky Mountains and Coast Ranges, being practically closed to the north about latitude 55° 30' by the ends of several rather high mountain ranges, which are here interpolated between the Rocky Mountains and Coast Ranges. Nearly coincident with the forty-ninth parallel, it is cut off in the same manner by a second transverse mountainous region, formed in a similar way to the first, which separates it from the great lava-plain of the Columbia River. It has thus a length of about five hundred miles. Its elevation decreases gradually from south-east to north-west, and it is now traversed in various directions by systems of trough-like valleys of erosion, which are deepest where the general level of the plateau is greatest. Water standing at an elevation of 3,000 feet above the present sea-level would flood most of these valleys in its southern part, while in its northern portion, a large tract of country about the fifty-third and fifty-fourth parallels of latitude would be completely submerged by it. The mean elevation above stated must therefore be accepted merely in a general way, and is an approximation to the average height which this region might have if its irregularities were levelled down. It is in fact, as a rule, only when broadly viewed, and in contrast with its bordering mountain ranges, that its true character as a plateau becomes apparent. North of the Cariboo Mountains on the fifty-fourth and fifty-fifth parallels of latitude, the plateau region interrupts the line of the Gold Ranges and abuts directly on the inner slope of the Rocky Mountains proper. Southward from the Cariboo Mountains, it is irregularly bounded to the eastward by the several members of the Gold Ranges.

Beyond the northern end of the Interior Plateau, the interval between the Rocky Mountains and Coast Ranges appears to be occupied by an irregular mountainous country about which little is yet known, for a distance of about 250 miles, till, in the vicinity of the fifty-eighth parallel, this country again begins to assume a plateau-like character, having at first a height of about 2,500 feet, but sloping down gradually north-westward, and constituting the upper drainage basin of the Yukon.

Having thus very briefly characterized the main features of this western portion of the continent as they now exist, we are in a position to follow in greater detail the steps by which they have been produced. Omitting then from consideration the imperfectly known progress of events in the earlier stages of the geological history of the region, we may endeavour to picture to ourselves its condition in the Triassic or first stage in the

Mesozoic division of geological time. The central region of the continent was at this time occupied by a very extensive, though shallow, mediterranean sea, which was either entirely cut off from the ocean or had only occasional and brief connection with it, and in which red beds with occasional layers of gypsum and salt were being deposited.[1] Rocks which represent a portion of the bed of this inland sea enter into the composition of the Rocky Mountain Range near the forty-ninth parallel, but are not known to occur to the north of that parallel for a distance of more than thirty or forty miles. To the west, they are not found in the Selkirk or Purcell Mountains. We appear in fact to discover in this vicinity the northern end of the inland Triassic sea. To the west of the Gold Ranges (under which term it will be remembered that the Selkirk, Purcell and other mountains are grouped), deposits also referable to the Triassic period, and more particularly to its upper part, are again found. These occur both on the mainland of what is now British Columbia and on Vancouver Island and the Queen Charlotte Islands. They contain truly marine fossils, and consist largely of materials of volcanic origin, which give evidence of contemporaneous volcanic activity on a great scale. To the north, in the Peace River country, and to the east of the present position of the Rocky Mountains, rocks holding the same marine forms are found, and they have quite recently again been discovered by Mr. McConnell in a similar position, still further north, on the Liard River.

It would thus appear that in Triassic times the eastern border of the Pacific washed the western slopes of the Gold Ranges, and that where this mountain-system became interrupted, in its northern part, the sea was continued across its line and covered a large tract of country to the east of the present position of the whole Cordillera belt.

Precisely how far to the east the shore of this northern expansion of the Pacific was situated, has not yet been determined. The region between it and the northern end of the inland sea previously referred to must have been a land area, which separated the open ocean of the north from the mediterranean on the south. The Rocky Mountains proper had not yet been formed, nor is there any evidence of mountain ranges in the region of the Coast and Vancouver systems of to-day, though the volcanic action there in progress may have produced insular volcanic peaks. The deposits of the inland Triassic sea, including as they do beds of salt and gypsum, appear to prove the existence of a very dry climate in the area occupied by it, and as the land barrier separating it from the moisture-bearing westerly winds of the Pacific can not have been wide, it must have been high. It is thus probable that the mountains of the Gold system formed at this time a lofty sierra, which was continued to the south of the forty-ninth parallel by the Cabinet, Cœur D'Alaine, Bitter Root and other mountains at least as far as the Wahsatch Range in Utah.

The Triassic period was closed by one of those epochs of folding and dislocation of strata which are found to be recurrent in geological time, and which are generally attributed to the secular contraction of the earth's crust. The evidence of this time of change has been examined in greatest detail in the vicinity of the present coast-line, where it resulted apparently in outlining the Vancouver and Coast Ranges, and was accompanied by the production or extravasation of great masses of granitic rocks.[2] It is highly probable

---

[1] Cf. "Note on the Triassic of the Rocky Mountains and British Columbia." 'Transactions Royal Society Canada,' vol. i, Section i, p. 143.

[2] Cf. 'Report of Progress Geological Survey of Canada,' 1878-79, pp. 46 B, 48 B; 'Annual Report Geological Survey of Canada,' 1886, p. 15 B.

that some corrugation along the line of the Rocky Mountains occurred at the same period, as, in the earlier Cretaceous strata next succeeding, without further evidence of disturbance, conglomerates are found composed of fragments of many varieties of the older rocks, which could scarcely otherwise have been rendered subject to denudation. Though much remains to be discovered respecting this post-Triassic epoch of disturbance, it was evidently an important one, and its results were widespread in the Cordilleran region. It is quite possible that it was accompanied by, or resulted in producing, a general elevation of this entire region above the sea-level, as no rocks certainly referable to the Jurassic or next succeeding period have yet been distinctly recognized either in British Columbia or in its bordering regions.[1] It must be borne in mind, however, that a portion of the red beds of the inland sea, described as Triassic, may extend upward into the Jurassic period, and that the marine Triassic fossils of the western and northern sea are referable to the later stages of the Triassic, or "Alpine Trias" of the Cordilleran region, comparable with the St. Cassian and Hallstadt beds of the Alps in Europe ; while the beds of the Cretaceous next found are, according to European analogies, near the base of that formation. As, moreover, it is far from certain that the occurrence of similar organic forms in regions so widely separated is satisfactory evidence of precise synchronism, it is quite possible that the unrepresented interval here is not very great. That, in other words, the highest beds attributed to the Triassic and the lowest of those given to the Cretaceous may overlap on both sides the period which, judging from European analogies, might be supposed to intervene, and which is elsewhere represented by several successive and more or less distinct faunas. On the other hand, there is every reason to believe, that those mountains at least which had been formed near the coast during the epoch of disturbance above referred to, had, before the earliest Cretaceous beds were laid down, been greatly reduced in importance, if not almost completely removed by denudation. The balance of evidence, indeed, appears to be in favor of the belief, that for a somewhat prolonged period succeeding that of disturbance at the close of the Triassic, this part of the Cordilleran region was throughout a land area, the western margin of which lay some distance beyond the present position of the country which now constitutes the sea border.[2]

The next distinct record of the physical conditions of the region under discussion is afforded by the earlier Cretaceous rocks. These, on the evidence of contained molluscan fossils, are regarded as about equivalent to the Gault of England, though the associated remains of plants are such as to admit their assignment to a somewhat older date. At this time, the immediately post-Triassic elevation had been followed by a subsidence of the land, resulting in the re-occupation by the open sea of the great area which had been similarly characterized in the Triassic. As in Triassic times, we find that this earlier Cretaceous extension of the Pacific, to the north of the fifty-fourth parallel, spread eastward

---

[1] Certain rocks, from which fossils supposed at the time to be Jurassic were described, have since been found to belong to the earlier Cretaceous. Cf. 'Report of Progress Geological Survey of Canada,' 1876-77, p. 150 ; " Mesozoic Fossils," vol. i, p. 258.

[2] It must be remembered that the successive foldings and crushings which the Cordilleran region has suffered, have resulted in an actual change in position of the rocks now composing its westward margin. This change may have amounted, since the beginning of Mesozoic time, to one-third of its whole present width, which would place the line of the Coast Ranges about two degrees of longitude further west. See in this connection remarks by Mr. R. G. McConnell, respecting the Rocky Mountain Range proper, 'Annual Report Geol. Surv. Can.,' 1886, p. 31 D.

in a more or less connected manner completely across the present position of the Cordillera belt, while the Gold Ranges, and probably also many other insular areas, continued to exist as dry land. In this case, as in that of the Triassic, it has not yet been found possible to outline exactly the eastern limit of the sea, in consequence of the want of sections cutting down to the base of the Cretaceous in the area of the Great Plains. There are, however, reasons for believing that it did not extend far beyond the line of the present foot-hills of the Rocky Mountains.[1]

In one important particular, the conditions in this earlier Cretaceous period differed from those of the Triassic. There was at this time no isolated inland sea, and waters in connection with the main ocean stretched southward to the east of the Gold Ranges as far as the forty-ninth parallel, and beyond it to a further distance which is as yet undetermined. This extension of the open sea thus actually overlapped, to a considerable extent, the area formerly occupied by the Triassic mediterranean.

No important beds of pure limestone or other evidences of deep sea conditions are found in these earlier Cretaceous beds, and there is on the contrary abundant proof of shallow waters, and occasionally of the local existence of tracts of low land, upon which a luxuriant vegetation existed, producing in some places important beds of coal. As, moreover, these local terrestrial conditions are recurrent throughout a great thickness of strata, it is obvious that the subsidence just referred to was continuous, or nearly so, and was followed *pari passu* by sedimentation.

About the stage in the Cretaceous which is represented by the Dakota group, however, a much more rapid downward movement of the land occurred. This is marked by the occurrence of massive conglomerates, which have now been recognized in many places in the southern part of the interior of British Columbia, as well as westward to the Queen Charlotte Islands, northward to the upper waters of the Yukon and Porcupine, and eastward along the line of the Rocky Mountains. This subsidence was not only more rapid but also more extensive than that which had previously been in progress. It was at this time that the open sea first spread across the entire area of the Great Plains, overlapping even the western flanks of the Laurentian plateau, and flowing southward as a wide strait to the Gulf of Mexico.[2] At or about this stage of the Cretaceous, important evidences of renewed volcanic activity are found both in the Queen Charlotte Islands and in the Rocky Mountain Range, and also probably in several places in the vicinity of the Coast Ranges of British Columbia.[3]

The central part of the Cordillera belt, however, for a great part of its entire length, still formed a more or less continuous land-barrier between this great strait and the western ocean. It is only in the extreme north-west that the waters of these two areas are known to have been freely in communication across it, and there is next found a considerable body of evidence (though for the most part of a negative character),[1] to show

---

[1] See map No. 1. This and the other maps attempt only to show the conditions at the times to which they are referred, in the most general way, no sufficient data having yet been obtained, for detailed delineation.

[2] For further detail see a paper by the writer on the "Earlier Cretaceous rocks of the North-western portion of the Dominion of Canada." 'American Journal of Science,' III, vol. xxxviii, p. 120.

[3] 'Report of Progress Geol. Surv. Can.,' 1878-79, p. 69 B.; 'Annual Report Geol. Surv. Can.,' 1885, p. 106 B.; 'Amer. Journ. Science,' III, vol. xxxvii, p. 127; 'Geological Magazine,' Dec. II, vol. viii, p. 218.

[4] Consisting chiefly in the absence of the next succeeding beds of the Cretaceous in the area outlined below. Evidence of folding and erosion at about this time has been found to the south in the Western States. *Cf.* "S. F. Emmons," 'Bull. Geol. Soc. Am.,' vol. i, pp. 276-279.

that not very long after the great subsidence, the barrier above referred to was still further elevated, and increased in width to the westward, till it included almost the entire area of the southern mainland of British Columbia, together with the southern part of Vancouver Island and the Puget Sound region.[1] It is further clear, that while subsidence continued, or the effect of the main subsidence had not yet been reversed in the eastern portion of the area of the Canadian Great Plains, that toward the west and north-west of the area now occupied by these plains, either some local and irregular elevation of portions of the sea-bed occurred, or sedimentation went on so rapidly as to more or less completely fill portions of the great Cretaceous strait. This is shown by the occurrence of estuarine, lacustrine and terrestrial conditions in the Belly River and Dunvegan series of different parts of that region. Next, as evidenced by the ubiquitous occurrence to the east of the Cordillera of the marine Pierre shales, there happened a further subsidence, which continued progressively or stage by stage to and throughout the Laramie or closing period of the Cretaceous.

Concurrently with these later Cretaceous conditions to the east of the Cordillera, it appears that on what now constitutes the littoral of British Columbia, to the west of the axis of the Coast Ranges, the area of Cretaceous sedimentation was transgressively extending to the southward, the local base of the Cretaceous being found at successively higher stages in the system in that direction, till at a time which is believed to have corresponded with the Laramie of the plains, the sea invaded the Puget Sound region.[2]

Thus, at the closing epoch of the Cretaceous period,[3] on the western margin of the Cordilleran region, massive accumulations of strata of a delta-like character were in process of formation in the Puget Sound region,[4] and very probably also as far as the northern end of the present Strait of Georgia; the land being along this part of the littoral somewhat lower relatively to the sea-level than now, and the source of the detritus brought down by rivers being found in the elevated interior of British Columbia.[5] To the north, in about latitude 61°, in the central region of the Cordillera belt, where the upper waters of the Yukon are now found at an elevation of 2,000 to 3,000 feet, evidence is again obtained of the presence of the sea or of an estuarine body of water at approximately the level of the ocean. To the east of the Cordillera, in lacustrine basins of great extent, which were in more or less free connection with the sea, and extending nearly to the line of the present Rocky Mountain Ranges, if not in places overlapping this line, the latest beds of the Laramie were in course of deposition.

There is, further, reason to believe that at this time a nascent Rocky Mountain Range was in part marked out by low hills, in which some of the Palæozoic and Triassic beds were already subject to subaërial denudation, and that rivers rising on the flanks of the Gold Ranges and flowing across the line of the present Rocky Mountains may have supplied a portion of the material of the Laramie of the North Saskatchewan district, while further south a great part of the material came directly from the prototypal representatives of

---

[1] Cf. for the corresponding southern region, S. F. Emmons, 'Bull. Geol. Soc. Am.,' vol. 1, p. 278.
[2] Cf. 'American Journal of Science,' III., vol. xxxix., p. 182.
[3] The Laramie is, for the purposes of the present discussion, assumed to represent the highest formation of the Cretaceous, as, though its flora leads to the belief that it comprises passage beds to the earliest Tertiary, it is physically attached to the Cretaceous rocks.
[4] Dr. C. A. White, 'Bulletin, United States Geol. Surv.' No. 51.
[5] 'American Journal of Science,' III, vol. xxxix, p. 183.

the Rocky Mountains proper.[1] An area including the entire inland portion of British Columbia, which is marked out by the borders of these regions of deposition, has afforded no evidence of Laramie rocks, and there is every reason to believe that it constituted a somewhat elevated tract of land at this time.[2]

This state of affairs was brought to a close by another of the recurrent epochs of folding and dislocation of the earth's crust, which was one of the greatest of those of which we find the results in the region under discussion, as well as the last of an important character to which it was subjected. Under the influence of enormous pressure acting from the Pacific side, the strata, till then nearly horizontal, which bordered the Gold Ranges on the north-east, were folded together and thrown up into a dominant ridge of Alps, which finally outlined the Cordilleran belt on this side. A similar folding and upthrust affected also the western marginal mountains which have been referred to as the Vancouver system, but the action was there probably less violent and certainly affected a narrower zone. A portion of the crumpling to which the rocks of the Coast Range have been subjected was doubtless also produced at or about the same time, and certain granitic extrusions which cut the earlier Cretaceous rocks on its eastern flanks, as well as much of the flexure of these Cretaceous rocks, are further attributed to this period of disturbance.

There is really no means of ascertaining what effect this dynamic movement produced in the region of the Gold Ranges, but it is more than probable that the whole width of the Cordillera then suffered change and deformation of such a character that little if any trace of its surface contour of an older date can be found to-day.[3] It does not, however, necessarily follow that the general altitude of the Cordillera belt was at this time greatly changed. The greater part of the accumulated pressure appears to have been relieved by folding along the lines of its two bordering ranges, and it seems to be not improbable, as a general proposition, that changes in elevation affecting wide areas are due to other causes than those producing mountain ranges.[4] We are warranted in assuming, however, that a certain movement in elevation was coincident or nearly so with that of the great disturbances above outlined, as no strata representative of the Eocene period proper have yet been found anywhere in the western part of Canada. The entire area of the Great Plains was thus sufficiently elevated to become dry land, as occurred at the same time in the Western States to the south of the international boundary.[5]

---

[1] Cf. 'Report of Progress Geol. Surv. Can.,' 1882-84, p. 113 C; 'Annual Report Geol. Surv. Can.,' 1886 p. 135 E.

[2] It may be noted, that to the south of the international boundary, in Washington, Dr. C. A. White and Mr. Bailey Willis find reason to suggest that certain beds occurring to the east of the Cascade Mountains, are of the same age as those of the Puget Group.—'Bulletin, U. S. Geol. Surv.,' No. 51, p. 54. See also S. F. Emmons, 'Bull. Geol. Soc. Am.,' vol. i, p. 282.

[3] In respect to this great epoch of orographic movement, as evidenced particularly in the more southern part of the Cordillera, which has now been somewhat closely studied, Mr. S. F. Emmons may be quoted as follows:—"It is unquestionably one of the most important events in the orographic history of the entire Cordilleran system. With the exception of the great unconformity between the Archæan and all overlying sediments, which is a phenomenon sui generis and altogether exceptional, no movement has left such definite evidence as that which follows the deposition of the coal-bearing rocks, to which the name Laramie has by universal consent been applied."—' Bulletin Geol. Soc. Amer.,' vol. i, p. 283.

[4] Cf. LeConte, 'American Journal of Science,' III, vol. xxxii, p. 178.

[5] It should be mentioned, however, as exceptions to this general statement, that a small occurrence of beds believed to represent the Green River Eocene, has been noted by Dr. C. A. White near the forty-seventh parallel

In the fortieth parallel region, during the Eocene period, many thousand feet of beds holding characteristic fossils were laid down in a series of lakes between the Rocky Mountains and the Sierra Nevada, but no such deposits have been met with in any part of the northern Cordillera. The whole sweep of country from the Laurentian region, possibly quite to the now submerged edge of the Continental Plateau on the Pacific side, thus became, and continued to be throughout the earliest Tertiary, an area of denudation, within which, if any small areas of deposition occurred, the beds formed in these have either been subsequently removed or have become concealed by later deposits.

For this first period of the Tertiary proper, included between the mountain-forming epoch which closed the Laramie and the oldest beds of the Miocene or middle Tertiary of the West, we must therefore endeavour to discover traces of another kind, viz., those impressed on the land by sub-aerial waste and the erosion of rivers. Evidence of very prolonged and important action of this nature is believed to exist. It is quite probable as a consequence of the greater elevation of the land and the ridging up of the central parts of the Cordillera brought about by the great post-Laramie dynamic movement, that the excavation of the remarkable system of valleys began, which now in a partially submerged condition exist as fiords along the western margin of this part of the continent. No evidence can, however, yet be advanced on this point. The main result, which there is good reason to refer to this time of denudation, in that part of British Columbia which lies between the Coast and Gold Ranges, is the formation of a first Interior Plateau or plain,¹ of which extensive, though now more or less disconnected fragments still exist. These remnants of an old Eocene land-surface are most marked in the southern half of the Interior Plateau of to-day, where they now have elevations of from 4,000 to 6,000 feet above the actual sea-level, and stand 3,000 to 5,000 feet above adjacent valleys and low tracts of later origin. The highest portions of this older plateau are found attached to the flanks of the bordering mountain ranges,² and its general elevation decreases toward the central line of the present Interior Plateau, as well as gradually in the direction of that line to the north-west, till, in its northern part, its average elevation is not more than about 3,000 feet, and the actual valleys are cut much less deeply into its surface. Climbing to the level of this old plateau, or to that of some slightly more elevated point about the fiftieth or fifty-first parallel of latitude, the deep valleys of modern rivers with other low tracts are lost sight of, and the eye appears to range across an unbroken or but slightly diversified plain, which, on a clear day, may be observed to be bounded to the north-east, south-west and south by mountain ranges with rugged forms; and above which in a few places isolated higher points rise, either as outstanding monuments of the denudation by which the plateau was produced, or as accumulations due to volcanic action of the Miocene or middle Tertiary period.

Extending our search for evidences of the prolonged Eocene denudation still further north, we find a notable width of plateau, based on the denuded surfaces of Palæozoic and

---

of latitude on the line between Dakota and Montana. Also that, according to Emmons, certain beds lately found by Mr. R. C. Hills at Huerfano Park, on the eastern flanks of the Rocky Mountains, are classed as Eocene—'American Journal of Science,' III, vol. xxv, p. 411. 'Bulletin Geol. Soc. Amer.,' vol. i, p. 286.

¹ Following Mr. W. M. Davis, the proximately level denudation-surface thus formed may be named a *peneplain*.—'American Journal of Science,' III, vol. xxxvii, p. 430.

² It appears even to run back into the areas of some of these ranges, at an elevation of about 6,000 feet above the present sea-level. Cf. 'Annual Report Geol. Surv. Can.,' vol. iv, p. 21 B.

Mesozoic rocks in the country about Dease, Frances and Finlayson Lakes, now drained by streams at lower levels which discharge to the Pacific, or to the Mackenzie and the Yukon.[1] This has an elevation above the present sea-level of 2,900 to 3,200 feet, and may with considerable probability be assigned to this time. During the same period the great Columbia-Kootanie Valley which separates the Gold and Rocky Mountain Ranges (and to which reference will again be made) as well as the Flathead Valley, in both of which deposits believed to be Miocene occur, must have been cut out by rivers flowing to the southward.[2] It is further probable that many more of the larger valleys of the more mountainous tracts also date from the Eocene period, even when these are actually known to contain Miocene beds.

It must be explained here, in advance of such remarks as will subsequently be made respecting the Miocene period, and to meet objections which may possibly occur to the reader as to the permanence as a table-land of a denudation-surface of such great age, that although the beds attributed to the Miocene are often found to be inclined at considerable angles, thus indicating disturbances subsequent to the time to which the main features of the old peneplain are attributed, such disturbance is usually local, and is probably confined for the most part to the vicinity of foci or lines of actual Miocene volcanic action; while the horizontal or nearly horizontal character of most of the Miocene volcanic rocks, in itself affords evidence of the general stability and undisturbed character of the old surface. It may be added, that while a considerable part of the area of this old peneplain is now capped by basaltic and other accumulations of the Miocene, such capping is, over at least eight-tenths of its area, too inconsiderable in thickness to be regarded as of importance in its bearing on the general question.

A plain like that of which the remnants are here found, and based as it is in various places upon granites and folded and disturbed Palæozoic and earlier Mesozoic rocks for the most part highly indurated (though varying considerably in this respect), can only have been produced, as I conceive, in two possible ways, viz., (1) as a plain of marine denudation, or (2) by a system of streams and rivers cutting down and working over a land-surface during a vast lapse of time, and under conditions of great stability, till that surface has become throughout approximately worn down to the base-level of erosion.

In this case, the first method is inapplicable, on account of the confined character of the area affected, which because of its bordering mountain ranges could have been in no direction open to the ocean, and is further negatived by the absence of strata, whether marine or otherwise, not only in the area itself, but also in the region to the east and along the border of the Pacific to the west of it, referable to the period in which the plateau is believed to have been formed. It is therefore to the second cause that the production of this early Tertiary peneplain must be attributed.

The circumstances which resulted in the production of this old nearly level surface I believe to have been as follows:—At the close of the great post-Laramie disturbance, the form of the surface of the Cordillera was such as to give rise to a river system, the limits of whose drainage basin nearly coincided with those of the Interior Plateau of to-day. It would be unsafe, from the existing slopes of the remaining parts of the old surface, to draw any conclusion as to the direction of outflow of this river-system, but the occurrence

---

[1] 'Annual Report Geol. Surv. Can.,' 1887-88, p. 109 B.
[2] 'Annual Report Geol. Surv. Can.,' 1885, p. 30 B.

of a high mountain barrier near the forty-ninth parallel tends to show that it discharged in a northerly direction. Whether its drainage issued eventually to the west, across the line of the Coast Ranges, or to the east across that of the Rocky Mountains, there is no available evidence to show. Little by little this river and its tributary streams, aided by other subaërial agencies, cut down almost its entire drainage-basin, till this became a nearly uniform plain, with some slight slope in the main direction of the river's flow; but of which the lowest part approximately coincided with the sea-level of the time. The lower part of this plain having to-day an elevation of about 3,000 feet above the sea, indicates that at the time at which this long continued action came to an end, the level of the central zone of the Cordillera stood lower by about that amount than it now does. After reaching this base-level of erosion, the rivers would of course be unable to do more than serve as channels for the conveyance of material brought into them from the surrounding country, which, wherever it stood above the general level, was still subject to waste. The valleys became wide and shallow, and the surface as a whole assumed permanent characters.

Round the borders of the region in which a peneplain was thus formed, higher parts of the original surface remained, though in much reduced form, as rugged mountain-partings between this river-system and others on the east, south and west. It is possible that some movement in elevation still continued along these mountain axes, which did not affect the Interior Plateau region, or affected it but little. This, however, is merely a conjecture which may help to explain the dominance which these mountains evidently continued to maintain.

Having reached the relatively stable condition of a peneplain, approximately at the base-level of erosion, the surface of what is now the Interior Plateau region may have continued to exist for a long time without notable further change; but this state of affairs was eventually brought to a close by some further orographic movement. According to Mr. King,[1] several such movements took place in the Cordilleran region of the fortieth parallel during the Eocene, and one of particular importance is attributed to the close of this period. There is, in British Columbia, some reason to suspect that a considerable amount of erosion of the surface of the Eocene peneplain occurred before the initiation of the deposition of the Miocene. This might be explained by a moderate elevation effected before the close of the Eocene; but it was probably by a more important movement, which took the form of a re-elevation of the mountain axes, and which may have been synchronous or nearly so with King's post-Eocene disturbance, that the Miocene lacustrine conditions were brought about. By an interruption of the drainage produced in some such way, the great Miocene lakes of that portion of British Columbia between the Coast and Gold Ranges were first formed. As to whether the disturbance referred to was accompanied by faulting along the bases of these bordering ranges, or by any considerable increase in the elevation of the Cordillera as a whole, no means has yet been found of determining; but the character of the Miocene flora seems to indicate that the land did not stand at this time at any very great height, and that the climate was temperate.

In whatever manner produced, there is ample and good evidence respecting the condition of the entire region here treated of in the earlier part of the Miocene. The

---

[1] "Geological Exploration of the Fortieth Parallel," vol. I, p. 765.

Interior Plateau became the site of a great lake, or more probably of a series of lakes of greater or less dimensions, some of which were of earlier date than others. That the region was not simultaneously covered by a single lake seems probable from the diverse lithological characters of the sediments met with, as well as from the different facies of the contained fossils, which practically consist, so far as yet observed, of plants and insects only. That some at least of these lakes were of considerable dimensions is indicated by the massive shingly accumulations of certain localities. As far north-west as the Francis River, and quite beyond the limit of the Interior Plateau as previously defined, deposits which are referred to the Miocene have been found, and beds which are believed to belong to the same stage occur again on the Porcupine branch of the Yukon. Lakes which are referred with some probability to the same Miocene period also existed in part of the Columbia-Kootanie Valley and in that of the Flathead River, but no definite palæontological evidence of the age of these has so far been obtained.

During this lacustrine age the higher mountain ranges stood, as they did during the foregoing period, above the general level, and some portions also of the Eocene peneplain were probably unsubmerged at this time. A luxuriant warm-temperate flora flourished on these lands, and its remains have been included in the sediments of the lakes. Along the marshy borders of the water vegetable accumulations, now found as lignites and coals, were produced; insects of species long since extinct, sported in the sun; and land animals doubtless roamed through the forests, though in this particular part of the Cordillera no relics of these have yet been discovered.

On the littoral, to the west of the Coast Ranges, beds are found in a number of places which have been referred, on evidence more or less complete, to the same period; some being marine and some of lacustrine origin. In the Western States, the Pah-ute Lake of King was about this time formed to the east of the Sierra Nevada, while to the east of the Rocky Mountains the Great Plains suffered a depression, deepest along the Rocky Mountain foot-hills, which produced the Sioux Lake of the same author. The depression which gave rise to this last-mentioned lake probably extended northward into the Canadian plains, and was there greatest along a line running north-westward, parallel to the Cordillera, but at a distance of nearly 200 miles from its base. In the Cypress Hills and Hand Hills, outliers of Miocene rocks are found, upon which the statement here made is based, and these contain mammalian remains in some abundance, which are according to Prof. Cope referable to the 'White River period.' It is important to note that the beds here met with consist largely of conglomerates, the well-rounded pebbles of which have been derived from the harder beds of the Rocky Mountains, as the occurrence of these coarse materials at such a distance from their source implies the existence at the time of a considerable gradient from the base of the mountains to this line of greatest depression, a gradient such as to produce rapid rivers with great powers of transport, and involves besides the recognition of the fact that the western margin of the plains stood at a relatively high elevation.

Had the orographic conditions remained permanent for a prolonged period, the lacustrine phase of the Interior Plateau of British Columbia would have been in the end terminated by the filling up of some lake-basins, and the drainage, by the gradual cutting

---

[1] 'Annual Report Geol. Surv. Can.,' 1885, pp. 68 C, 79 C; 'Annual Report Geol. Surv. Can.,' 1880, p. 138 E; 'American Naturalist,' vol. xix, p. 163.

down of the beds of the effluent rivers, of others. Before this had taken place, however, a new factor appeared. Volcanic action was recommenced on a great scale; fragmental volcanic ejectamenta, supplied from numerous vents, were discharged and mingled themselves with the ordinary detrital deposits of the lakes, while basaltic and other lavas flowed out in great volume over large parts of the plateau country. The principal centres of this volcanic action appear to have been aligned near the eastern or inner base of the Coast Ranges, where the more or less degraded bases of some of the old volcanic vents may still be traced.[1] Isolated patches of volcanic material, which are probably due to the same period, are found, however, to occur far north, on the Stikine and in the Upper Yukon basin, as well as along the littoral of what is now the Province of British Columbia. It is probable that we may refer the volcanic activity of which evidence is afforded in one place near the southern boundary of the Canadian Great Plains to the same period. The Three Buttes, or Sweet Grass Hills, there rise as isolated monuments of such action, standing at the present day, by reason of the superior induration of their rocks, high above the level Cretaceous plain. These peaks are evidently now but the remnants of the necks of volcanic mountains, of which the history has not yet been traced in detail.[2] We shall have occasion to allude to them again in a later part of this paper (pp. 60, 67.)

Prof. LeConte believes that the great lava-flow of Middle California occurred near the end of the Pliocene, while similar irruptions to the north, in Oregon, took place somewhat earlier, at the close of the Miocene or early in the Pliocene.[3] The last named region closely corresponds with the Interior Plateau of British Columbia, but in British Columbia the evidences of the blending of the Miocene lake deposits with those due to volcanic action, are in some places so distinct, that we are amply justified in attributing at least the first eruptions to the Miocene period itself. That volcanic action may have continued into the Pliocene period can not, however, be denied, particularly in view of the impossibility, in the absence of physical changes, of drawing any perfectly distinct line between the various subdivisions of the Tertiary. All that can be said at present is, that there is no known proof of Pliocene volcanic activity in British Columbia.[1]

Taken as a whole therefore, the Miocene period, in regard to the Interior Plateau of British Columbia, may be described as one of deposition and accumulation of material. The then lower parts of the old Eocene peneplain were in the first place partially filled with lacustrine sediments, in the second more or less uniformly overspread by lava-flows or other volcanic accumulations. This levelling-up process was evidently so complete as entirely to obliterate the valleys of the Eocene and Miocene river-systems, and it is doubtful whether any one of these valleys has since been re-excavated along its old course for any considerable distance. This, however, applies only to the area of the Interior Plateau. Valleys formed in the more mountainous parts of the Cordillera have doubtless, in many

---

[1] 'Report of Progress Geol. Surv. Can.,' 1876-77, p. 75. Further, and as yet unpublished, evidence, on this point has since been found in the southern part of British Columbia.

[2] Cf. 'Report of Progress Geol. Surv. Can.,' 1882-84, p. 45 C.

[3] 'American Journal of Science,' III, vol. xix, p. 180; vol. xxxii, p. 177.

[4] Evidence is found, however, in some places, of which Pavilion Mountain and the upper part of Hat Creek may be specially mentioned, such as to show the existence of lakes or ponds of limited size in intervals of the period of volcanic eruptions. Some of these may yet afford organic remains of later date than the Miocene. The recurrence of volcanic phenomena during a considerable length of time is also shown by the cutting out of valleys in the basalts and the refilling of these by later basalts in the Stikine region. See 'Annual Report Geol. Surv. Can.,' 1887-88, p. 72 B.

cases, been since perpetuated, and the drainage marked out at this time has in such instances been maintained.

In the southern extension of the Cordilleran region, important orographic changes of various kinds, which were probably synchronous or nearly so with the epoch of volcanic eruption above referred to, took place at the close of the Miocene period. At this time the Coast Ranges of California were produced, the beds of the Pah-ute Miocene Lake were thrown into gentle folds, the area of the Great Basin sank, and became the site of an extensive lake. The plateau region of Utah was further elevated by an amount of from 2,000 to 3,000 feet, and a renewed gradual subsidence affected the Great Plains to the east of the Cordillera.[1] These movements appear to have been but faintly reflected in the region to the north of the forty-ninth parallel, but it is in all probability to this time that we must attribute the local folding of the Miocene rocks, which has been referred to on a previous page. There is, however, no valid evidence of any considerable change either by elevation or depression of the region of the Interior Plateau, where, at or shortly after the time at which the volcanic forces had exhausted themselves, the drainage began to outline a new system of stream- and river-valleys. Apart from probable local changes in level, which may alone have been sufficient to prevent the drainage from resuming its old direction, the filling up of the former valleys had been so complete, that the new streams, following the inclinations of the actual surface, began to cut their beds in courses entirely different from those of the old. Local accumulations of volcanic material, together with the slight folding to which the Miocene beds, whether volcanic or otherwise, had been subjected, must have caused the area of the old Eocene peneplain to be less uniform than before; and denudation acting on its projecting parts began again to reduce these toward the general level. This action was probably long continued, and had important effects in bringing the Interior Plateau region again down nearly to the base-level of erosion. The streams, which evidently flowed at this time with low gradients, cut out wide, shallow, trough-like valleys, which may yet be found in many places, and are at the present time in some instances still occupied by existing streams, though in others interrupted or abandoned owing to later changes, some of which will shortly be alluded to. These early post-Miocene valleys, where they may yet be studied, present all the characters of a drainage system which had been long maintained under stable conditions of the surface. They are found at the present day pursuing sinuous courses over the surface of the old table-land, and may also be recognized in many cases even in the mountain regions, where they appear as wide U-shaped valleys, in the bottoms of which winding streams pursue a comparatively tranquil course, and in which lakes, produced in various ways, are not uncommon.

At a later date in the Pliocene or last period of the Tertiary, on which we have now entered, it is evident that a very considerable and general elevation of this part of the Cordilleran region occurred. During the first stage of the Pliocene, the Cordillera must have been lower relatively to the sea-level than at present; when this elevation took place, it became considerably higher than it is now. The gradients of all the rivers were thereby increased, and being thus armed with new powers of erosion, the streams began to cut deep and, at first, narrow channels. To this time the cutting out of the greater part of

[1] King, "Geological Survey of the Fortieth Parallel," vol. I, p. 756; Le Conte, 'American Journal of Science,' III, vol. xxxii, p. 177; Dutton, "Tertiary History of the Grand Canon District," p. 226.

the deep valleys, which now exist in a submerged state as the fiords of the coast, is with all probability assigned. While admitting that these fiords may have been to some extent shaped and enlarged locally by ice during the Glacial period which followed, their depth is such as I believe to show that at the main period of their formation, in the Pliocene, the land stood relatively to the Pacific about 900 feet higher than it now does.

This later Pliocene elevation doubtless produced many important changes in the drainage system of the country, and particularly in that of the Interior Plateau region. Certain rivers which were fed from the perennial snows of extensive mountainous districts were then enabled, by the increased gradients given to them, to cut down so rapidly, compared to other streams which were less copiously supplied, as to achieve and maintain a dominant position, and even to completely rob some of the old valleys of their waters. It was doubtless at this time that the Fraser and its great tributary the Thompson managed to extend their drainage areas over so large a part of British Columbia. The Fraser may be described as running round the south end of the Coast Range of British Columbia, or, if regarded more strictly as cutting through the southern end of this mountain-system, it does so at a point of decreased elevation and width. Having in consequence been enabled in a comparatively short time to greatly deepen the lower part of its present course, this river next drew to itself the drainage of a great part of the northern and eastern portions of the Interior Plateau. Between the north and south part of the present Fraser valley and the inner borders of the Coast Ranges, from latitude 50° to latitude 54°, a broad belt of country seems at an earlier time to have been drained by several smaller rivers, each of which had cut for itself a valley completely through the Coast Ranges; but these rivers appear to have been gradually deprived of their upper tributary waters by the extension of the western tributaries of the Fraser, till, in the case of the Homatheo, and to a greater or less extent in that of other rivers, we find at the present day a very small stream making boldly toward the mountain range, and still carrying its waters through it in deep cañons to the sea, though apparently quite incapable of itself excavating a gorge such as that which it now occupies.

To the long continued denudation and erosion which were in progress during the Pliocene, the greater part of the deep pre-glacial auriferous gravels, of valleys such as those of the Cariboo Mountains, must be assigned, though it is at the same time possible that much of the excavation of these deep mountain valleys may have been more or less continuously in progress since the date of the post-Laramie disturbance. In the Interior Plateau region no deep auriferous gravels have yet been found, though it is more than probable that if the now concealed beds of the Eocene drainage system of this district could be discovered and followed, they also would be found to be characterized by important deposits of the precious metal.

It would be beyond the scope of the present sketch to endeavour to catalogue and describe the various older Pliocene river valleys which have afforded the evidence upon which the statements just made are largely based; but as the observations on these have not yet been published, it may be well to enumerate a few of them, and to point out the influence of the later Pliocene elevation of the land upon them. This influence is apparent chiefly in the formation of deep narrow gorges and cañons in the parts of these older and wider valleys which debouch upon the larger and lower river valleys, such as that of the Fraser, the Thompson and the Columbia. To keep pace with the rapidly deepening valleys of these main rivers, the streams of the tributary valleys were forced to cut down very fast toward their mouths, and this cutting down has gradually retrogressed (in many cases for a considerable distance) along the line of the old valley.

A few instances only of observed facts in this connection, derived from the southern part of the Interior Plateau region in the vicinity of the Thompson and Fraser Rivers, will here be cited. The heights placed after the names of streams in the subjoined tabular statement are those of the lower remaining parts of the older Pliocene valleys, below which, in each case, the present stream enters a narrow, steep, gorge-like valley, by which it descends to one of the large rivers or to an adjacent deep tributary valley of one of these rivers. While the main features of difference above noted as existing between the older and later Pliocene valleys are well marked, it is of course difficult in most cases to state the exact point at which the older valley should be considered to give place to the newer, and impossible to eliminate completely the influence of still later changes of greater or less importance.[1] The heights given must therefore be considered as approximate only. They are, however, it is believed, sufficient to illustrate and confirm the point in question:—

(Meadow Creek),[2] (upper valley)	3,350 feet
(Pukaist Creek), " "	3,750 "
(Witches' Brook) " "	3,650 "
Three-Lake Valley (south part)	2,750 "
Kelly Lake Creek (upper valley)	3,300 "
Cache Creek " "	2,050 "
(La-loo-wissin Creek), (upper valley)	3,450 "
Pavilion Creek	2,300 "

The difficulty above alluded to of eliminating the influence of other causes, both of earlier and later date than those of the Pliocene, renders it impossible to take the actually lowest instance of these or other similar wide low-grade valleys, as a definite point from which to measure the depth of the later Pliocene erosion. By striking an average of the height of the lower existing parts of a number of these old valleys, however, we may arrive at a rough approximation to the depth of this erosion, in that part of the region here particularly referred to. Such an average gives a height of about 3,200 feet above the present sea-level, while the mean level of adjacent parts of the Fraser and Thompson Rivers is about 730 feet. As it will be shown in the second part of this paper that these large rivers have not materially, if at all, reduced their general level since the close of the Pliocene, the difference between the two figures above given may be taken as roughly indicating the amount of cutting-down which they accomplished subsequent to, and as the result of, the later Pliocene movement of elevation, this difference being about 2,470 feet. The elevation of the land in this district, as compared with that in the earlier Pliocene time, must therefore be assumed to have considerably exceeded that amount.

Good illustration of the point here specially referred to, though in a more mountainous region, may be seen on the line of the Canadian Pacific Railway, in the Rocky Mountains

---

[1] Among which may be mentioned the partial infilling of many of these valleys with drift during the Glacial period.

[2] The names enclosed in parentheses have not yet appeared on published maps of the region. As the purpose here in view is merely that of illustration, it is considered unnecessary to explain in detail the positions of these streams.

and Selkirk Range.[1] In most of these valleys the wide and relatively level upper portions contrast markedly with their lower cañons and gorges. The Kicking-Horse, Upper Kootanie and Elk Valleys, while still wide, and above the narrow steep portions by which their rivers reach the adjacent part of the Columbia, have elevations of about 3,200 feet above the actual sea-level, and effect therefrom a rapid descent through newer and narrower valleys of from 700 to 900 feet to the Columbia.

These differences are much less than previously quoted, but it must be remembered that the localities referred to are much further inland, particularly if their distance from the coast be measured along the sinuous line of the Columbia River and not in a direct line from the sea. This, however, need cause no surprise, as even on the assumption of an equal elevation of the whole breadth of the Cordillera in the later Pliocene, the influence of such elevation on the upper parts of the larger rivers must have made itself manifest only after a long time, and must moreover, even if it was allowed time to produce its full result, have been much less effective at such a distance from the coast.

The best instance which has yet been studied of the relation now obtaining between the early Pliocene low-grade valleys and the more modern and deeper erosions which occurred during the later Pliocene period of greater elevation, is that afforded by what we may call the "Old Cache Creek Valley."

As I am enabled to present a map of this valley and its vicinity, a few words in explanation of its character may be admissible, as a conclusion to this first part of the general enquiry here made. I must, however, before going further, note that a detailed contoured map would much more adequately represent the actual conditions than that which is here given, on which the mode of delineation adopted does not fully express the general plateau-like character of the region depicted, but makes it appear more mountainous than it really is.

The main line of the Old Cache Creek Valley in question, I have been able to trace out and to travel along for a total distance of about thirty-eight miles. Its general course across the now broken surface of this part of the Interior Plateau is nearly east and west, an approximate parallelism being maintained to the direction of the corresponding part of the Thompson Valley. The total difference of level between the upper or eastern end of the remaining part of the valley and its western determinable end is approximately 530 feet, giving a slope of nearly fourteen feet to the mile in the direction in which its water originally flowed. The amount of the fall thus indicated, being nearly that which might be expected from the character of the old valley, in itself affords a valuable indication of the absence of any considerable differential elevation in this part of the Interior Plateau since early Pliocene times.

Standing on some prominent point of the plateau, one may trace the wide trough-like form of this old valley for many miles, trenching the surface of the plateau and pursuing a nearly direct westward course. The stream by which this notable valley was formed has, however, been unable to perpetuate itself. The history of its decline and interruption I conceive to be as follows :—

---

[1] I am aware that the occurrence of such U-shaped and wide valleys in mountain regions has often been attributed to the action of glaciers in such valleys. This explanation does not, however, require to be resorted to here, nor am I prepared to admit that it is a valid one.

The stream by which the valley was produced and maintained for a long period in the earlier part of the Pliocene was fed only by drainage gathered from the surface of the plateau itself, while the Thompson River, rising in the perennial snows of the mountains to the east, was much more bountifully supplied. So long as the surface of the plateau stood at a low elevation, this difference in character of supply as between the Old Cache Creek and the Thompson was not of vital importance to the first-named stream. The Thompson having soon cut down to the approximate base-level of erosion which was not much below that of the Old Cache Creek, had no perceptible effect on the latter. With the greatly increased elevation given to the region in later Pliocene times, however, all this was changed. The Thompson then began rapidly and persistently to deepen its channel, while the rival stream, with its small volume of water, was unable to do much more than to pursue the even tenor of its way. Very soon the small intervening lateral tributaries of the Thompson, in order to keep pace with the deepening of the main valley, began to cut deep narrow channels, the heads of which crept back farther and further from the Thompson, till, eventually, in several cases, they reached and tapped the drainage of the Old Cache Creek Valley. At about this point some of them ceased to cut back, but others, meeting with northern tributaries of the Old Cache Creek, quickly adopted the waters and the channels of these, till in the end a new series of important transverse valleys, all tributary to the Thompson, was produced, and a great part of the Old Cache Creek Valley was left high and (nearly) dry in the form of isolated troughs upon the table-land.

It will be unnecessary to enter into further details of this interesting history, which can be traced out on the map, upon which the course of the Old Cache Creek Valley is marked by the pink line. It will be noted that the small modern stream known as Cache Creek has, following the line of the old valley, cut back for a certain distance by a narrow recent gorge, but long before it could deepen the entire length of the old valley, other streams had largely deprived it of its erosive power by robbing it of its upper waters. Of these Eight-mile Creek, Deadman River and the Tranquille River and its tributaries are the chief offenders. It will further be observed that the branch of the Deadman known as Criss Creek is supposed to coincide for a considerable part of its length with a principal northern tributary of Old Cache Creek, which it has now for a number of miles cut down into the form of a nearly V-shaped and later valley. The course of the Tranquille is also identical for some miles with that of the old valley. With reference to the heights of various parts of the line of the old valley, it may be stated that the actual levels of summits in this valley between the intersecting valleys of the newer streams are shown in heavy figures. As, however, the old valley has been largely choked by drift deposits, due to a still later period, from which it has since been only partially cleared by denudation, the various heights given can be accepted only as approximately indicating the actual level of the several parts of the old valley-bottom.

From what has been stated in the foregoing pages, it will be evident that throughout the whole Cordilleran region of British Columbia, the prolonged era of denudation of the Pliocene produced very important results. The rivers and streams acted in the first place for a long time during a relatively low stage of at least the greater part of the land-surface, and subsequently for a similar period at a high stage. During this latter many of the first-formed valleys were greatly deepened in their courses as originally outlined, while others were robbed of their drainage and new valleys were produced with other directions. These results of the Pliocene erosion are of special importance from a geographical point of view, as the whole effect of subsequent events can be shown to have been relatively insignificant, and the main features of the surface of the country still remain much as they were at the close of the Pliocene period. The principal rivers and streams still, with small exceptions, occupy the same valleys, and since the Glacial epoch (as subsequently noted) many of them have been unable to clear their valleys from glacial debris down to the level of the later Pliocene channels. The old Eocene peneplain of the Interior Plateau, raised in the latter part of the Pliocene to a level considerably in excess of that which it now possesses, became divided by deep erosions into block-like masses, the edges of which were subsequently more or less reduced, while their central portions were but little affected. This reduction of the peripheral portions of the areas between streams must have continued more or less, however, to the present day, and has gone so far in many cases as to leave irregularly rounded high areas between the deeper valleys, of which only the upper portions are at about the general level of the old peneplain. Elsewhere rather extensive areas of nearly level plateau remain, and the denudation and erosion have seldom if ever been carried so far as to produce rugged forms such as those found in the higher mountain ranges.

While such effects were being produced in the inner parts of the Cordilleran belt, we may enquire as to the contemporaneous condition of the coast region of British Columbia on the west, and as to that of the Great Plains to the eastward. On the coast negative evidence only is available, but it appears probable that, as a result of the later Pliocene elevation, a belt of low land, wide enough to include Vancouver Island and the Queen Charlotte Islands, was formed, across which rivers fed by the streams issuing from the valleys of the Coast Ranges flowed to the sea. The Fraser doubtless flowed round the south end of Vancouver Island, receiving important tributaries from the valley now occupied by the Gulf of Georgia and from the region of Puget Sound.[1] A second river of some importance, carrying the drainage of the Coast Ranges to the north of Bute Inlet, probably discharged its waters to the north of Vancouver Island, while between the Queen Charlotte Islands and the mainland a wide plain must have been formed.

The Great Plains, to the east of the Cordilleran belt, are in the main plains of deposition, based for the most part on unaltered and undisturbed Cretaceous and Laramie strata, and only in a very secondary way owe any of their present regularity of surface to planing down by denudation. It has already been stated that, during the earlier portion at least of the Miocene period, deposition was going on over some part of the plains to the north of the forty-ninth parallel, as well as over a very extensive area to the south of that parallel in the Western States. Denudation must, however, have been in progress over other unsubmerged parts of the plains, and, to the north of the forty-ninth parallel, particularly in that steeply-sloping western portion which it has been shown must have existed between the central area of deposit and the corresponding part of the base of the Rocky Mountains. (See p. 14.)

In the Western States Mr. King describes the occurrence of a further general subsidence of much of the extent of the plains at the close of the Miocene, which resulted in the formation of a vast Pliocene lake, named by him the Cheyenne Lake, the sedi-

---

[1] *Cf.* Newberry in 'Annals of the New York Academy of Science,' vol. iii, p. 265.

ments of which enormously overlapped those of the previous Miocene lake in every direction,¹ but no trace of the extension of this lake over the Canadian Great Plains has been found. It is, on the contrary, probable that this part of the plains was during most of the Pliocene period in process of reduction by waste. The amount of this denudation can be approximately determined only in the vicinity of the Cypress Hills and Hand Hills, which, being capped by resistant Lower Miocene beds, now rise as isolated plateaux. In the first mentioned region the general surface of the adjacent plain is thus found to have been reduced by about 2,000 feet, in the second by about 1,000 feet.² These instances are sufficient to show that an enormous quantity of material must have been removed from the general surface of the plains since the earlier part of the Miocene period, and in all probability the greater part of this denudation occurred during the Pliocene.

It must, however, be added that certain deposits occur in the area of the Great Plains north of the forty-ninth parallel, to which a Pliocene age is assigned. These consist generally of beds of gravel, with some sandy and silty layers, which rest indifferently on various members of the Cretaceous and Laramie and lie beneath the lowest boulder-clay of the Pleistocene or Glacial epoch. These have now been found in a number of places throughout the districts of Alberta and Assiniboia, and have been named by Mr. McConnell the South Saskatchewan Gravels. Much of their material is evidently derived from the Miocene conglomerates, and this has been re-arranged at lower levels in the beds of rivers and in lakes of greater or less dimensions. These deposits, however, evidently belong to the very latest stage of the Pliocene, and are in fact assigned to this period only on the assumption that any beds formed before the onset of distinctly glacial conditions must be called Pliocene. In several places pebbles or small boulders of Laurentian origin have been found in the upper parts of these deposits both in Alberta and Assiniboia, implying transport from long distances to the east or north-east. Such transport, it is believed, can scarcely be accounted for by river action, and it appears to be more probable that we here find evidence of the beginning of that time of submergence of the northern part of the Great Plains, which, in accordance with views subsequently expressed, culminated during the Glacial period; and that when the upper beds of the South Saskatchewan Gravels were laid down, field-ice at least was already beginning to float westward or south-westward from the edges of the Laurentian plateau, and carrying with it, from beaches or the estuaries of rivers, specimens of the materials of which this plateau is composed.³

It would appear, from the facts here referred to, that the later or mid-Pliocene elevatory movement of the Cordilleran region in British Columbia either did not produce any effect on the adjacent area of the Great Plains, or that if it did so it was that of a correlative subsidence, or was shortly followed by an independent subsidence of a great part of these plains. Further south in the Cordillera, Prof. Le Conte finds reason to believe that, at the close of the Tertiary and inaugurating the Pleistocene,

---

¹ "U. S. Geol. Expl. Fortieth Parallel," vol. i, pp. 542, 756.
² 'Annual Report Geol. Surv. Can.,' 1885, p. 69 C.; 1886, p. 78 E.
³ For details respecting these latest Pliocene or early Glacial Deposits, see 'Report of Progress Geol. Surv. Can.,' 1882-84, p. 140 C. 'Annual Report Geol. Surv. Can.,' 1885, p. 70 C.; 'Annual Report Geol. Surv. Can.,' 1886, p. 139 E.

the elevation of the Sierra Nevada was considerably increased with the production of greater slopes on the west and of faulting along the east side of this range. The evidence consists chiefly in the existence of great river erosion, referred by him to this date, but he quotes further the observations of Dutton, Gilbert, Howell and Russell to show that other important orographic movements were synchronous, or nearly so, with the elevation of the Sierra. These comprise a general increase of the height of the Great Plateau of Utah of 3,000 to 4,000 feet and normal faulting in the region of the Great Basin and in southern Oregon, which has, in the last-named instance, affected the later Tertiary lava-beds.¹ The question now occurs as to whether these changes can be assumed to have been simultaneous with the mid-Pliocene elevation in the British Columbian portion of the length of the Cordillera. To this it may be replied that the great amount of erosion met with in British Columbia and referred to the latter half of the Pliocene period, implying as it does a long interval between the mid-Pliocene elevation there and the initiation of the glacial conditions, appears definitely to negative such correlation. In British Columbia no facts have yet been met with to prove a renewal of elevation at the close of the Pliocene, but it is at the same time possible that such elevation may have occurred, and, in view of the evidence brought forward by LeConte, it may even be said that it is quite probable that it did occur, and that, working together with other causes of a much more general character, it resulted in the initiation of the phenomena of the Glacial period. It is certain that the mid-Pliocene elevation was not in British Columbia sufficient in itself to bring about any general conditions of glaciation.

King refers also to the close of the Pliocene the wide change in relative levels which resulted in raising to greater elevations the western edge of the deposits of the Pliocene Cheyenne Lake, and the production of the eastward and southward slopes of this old lake-bed from the base of the Cordillera in the fortieth parallel region; but, so far as can be gathered from the published evidence of this great change, it can be asserted only to have been post-Pliocene. In the Canadian portion of the area of the Great Plains there is good evidence to show that the eastward or north-eastward slope of the actual surface was produced by movements during or at the close of the Glacial epoch.²

In reviewing the foregoing account of the changes in elevation of the Cordillera and its adjacent regions, including only that northern part of the western portion of the continent with which we have here to deal, the following general facts appear:—During the Laramie period a great part of the central region at least of the Cordillera stood at a high level, while the greater part of the area of the Great Plains was submerged. In Eocene times the plains ceased to be an area of deposition, and were somewhat elevated, the greater part of the Cordilleran belt being at the same time from 3,000 to 4,000 feet lower than at present. During the Miocene the Cordillera probably retained a nearly similar elevation, and there is evidence that the western edge of the plains stood high, but with a line of depression further to the east which ran parallel to the base of the Cordillera. The Pliocene was marked by two or more epochs of elevation of the Cordillera, while the plains, at the close at least of this period, are found to have been again depressed.

---

¹ 'American Journal of Science,' III, vol. xxxii, p. 167; "Tertiary History of the Grand Cañon," p. 227.
² This fact has already been noted in the 'Report of Progress Geol. Surv. Can.,' for 1882-84, p. 151 C.

In the above series of events it appears distinctly enough that the times of change in elevation of the provinces of the Cordillera and Great Plains respectively, though co-ordinated, did not correspond in character; while a general impression is conveyed that the movements were in the two areas complementary and more or less completely correlative. To this particular aspect it will be necessary to refer again in greater detail on a later page.

[25]

~~THE ROCKY MOUNTAIN REGION~~.

II.

GLACIAL HISTORY.

Having in the first part of this paper followed the orographic movements, the wider movements of elevation and depression, the periods of deposit and those of denudation in that northern part of the Cordillera to which our survey specially refers, through the Mesozoic and Tertiary eras, we are brought in contact with phenomena of a new kind, or at least of a kind of which no marked evidence had heretofore been found, those, namely, of the period or epoch of glaciation. More than any other changes which the northern hemisphere has undergone, these have in late years become a battle-ground of rival theories, in consequence of which the most extreme views have been maintained. While it is admitted on all hands that the Glacial epoch presents many problems of great difficulty, it would appear that much of the difference of opinion which has been developed has resulted from a disinclination to admit the application to this period of arguments and explanations similar to those which are freely allowed in other fields of geological reasoning. It would further appear that the complex character of the evidence met with, and the resulting difficulty found in embracing all the facts under a satisfactory hypothesis, seem to have driven many writers to take refuge in extreme theories of a simple character, under which very diverse observations have been included, even though some of these require to be led to a common centre by very devious routes. It is true that in meeting at this stage with evidence of ice action on a great scale, we are called upon to deal with a new agent capable of producing new effects; but it is also true that the concurrent action of various changes in relative elevation affecting oceanic and atmospheric currents, may have had much to do with the inauguration of the new state of affairs. Whether any still wider changes of a cosmic character co-operated with those above alluded to is a further question having little connection with our present enquiry, and one upon which nothing need be said here.

Regarded as the latest in the series of great changes which have been effective in bringing about the actual condition of the region under discussion, and being in consequence one of which the results are still everywhere apparent, the phenomena of this particular time might justify a very detailed mode of treatment and be made the subject of an extensive monograph. The object in view is, however, rather that of connecting the phenomena of the Glacial epoch with those of antecedent periods, referring for detail in respect to observed facts to previous publications. Many of the observations made by the writer in different parts of the Cordillera and in the region to the eastward extending to Lake Superior, have been separately published in a number of different papers and reports. This sketch may, however, afford a means of drawing these together and correlating them in a manner not heretofore attempted.

In the publications above referred to, it has been made a rule to detail the facts which have come under my notice in each particular district, and to suggest or discuss the deductions and more or less theoretical conclusions which these facts appeared to warrant. In the present instance this inductive method is necessarily reversed, in order to present the matter under consideration with clearness and brevity; this being the only mode available for a subject which depends on so great an accumulation of observations. With the object, however, of maintaining as close a connection as possible with the actual basis of our knowledge, numerous references are given to former publications, while certain observations which have recently been made, but which have not yet been published, are accorded a somewhat fuller treatment.

Though the present paper deals primarily with the Cordilleran region of Canada, it has been found essential to refer also to adjacent regions in the foregoing pages; and in connection with the present stage of the enquiry it is more than ever necessary to keep in view the conditions at the same time affecting the Great Plains. As, however, the glaciation of the Cordillera presents in itself a sufficiently complex problem, it will be best to consider this alone in the first instance, and subsequently, with the knowledge so obtained as a basis, to treat briefly of the contemporaneous phenomena to the eastward. This method is, moreover, rendered appropriate by the circumstance that while the glaciation of these two regions is without doubt correlative, the known evidences of the Glacial epoch in these two provinces barely overlap.

In a region with such pronounced physical features as that of the Cordillera, the solution of the problems offered by the traces remaining to us of the Glacial epoch are by no means so simple as in less rugged districts, and it is more than elsewhere necessary to keep clearly in view the chief outlines of the orography, as sketched on a previous page. It is thus not possible, without greatly exceeding the limits appropriate in this paper, to fully present the various local conditions which appeal to the eye in the field and at the time of observation. These circumstances render it difficult to do full justice to many of these observations, a difficulty which is increased by the non-existence of any really accurate detailed maps of large parts of the region.

All that is known of the Glacial period goes to show that, relatively to that occupied by the movements and periods of rest which we have previously examined, the whole time embraced by this epoch was short; while the results produced were such that, though striking enough because of their recency, they might almost, if not altogether, have been overlooked had they occurred at some long previous geological time. For the same reason the character of the data available for the history of this epoch differs considerably from that on which the geologist depends in the case of older deposits. Here we must appeal very largely to the nature and arrangement of incoherent deposits which still occupy the surface, while with regard to water-levels we may often directly consult beaches and terraces which still exist in a condition little changed from that in which they were produced.

The evidence brought forward in former pages presents to us at the close of the Pliocene the Cordilleran region at an elevation of at least nine hundred feet above that which it now has, and leaves it probable that by a further uplift at the end of the Pliocene period (and marking the close of that period in this region) this amount of elevation had been still further increased. The Great Plains were at the same time at a relatively low level, and had not had impressed on them that long gentle slope from the base of the Rocky Mountains to the east and north-east which they possess to-day. As nothing has been found to show that these two great areas of uplift and depression were separated by a line of faulting in the vicinity of the eastern margin of the Cordillera, it may be assumed that a hinge-like flexure occurred along this margin or not far to the eastward of it.[1]

The Cordilleran region, in consequence of its high elevation, and probably also in part as a result of other concurrent causes by which the northern hemisphere was affected at the inception of the period of glaciation, appears to have become at this time pre-eminently the condenser of the North Pacific. Precipitation occurred upon it chiefly in the form of snow, which was so much in excess of the influence of the summer heat as to accumulate from year to year. Great glaciers formed in the higher mountains, probably in the first instance among those situated nearest to the coast; but eventually the greater part of the region became covered and buried either in *névé* or beneath glacier-ice. The directions of motion of the glaciers at first produced were doubtless in conformity with that of the valleys of mountain streams, but at a later date, when the Cordillera became completely buried, a general movement was initiated from a region situated between the fifty-fifth and fifty-ninth parallels of north latitude, in south-easterly and north-westerly bearings.[2] The Cordillera, in fact, between the forty-eighth and sixty-third parallels, or for a length of about 1,200 miles, seems to have assumed an appearance closely analogous to that of Greenland at the present day, save that in consequence of the high bordering mountain ranges, with the general trend of these and of the lower intervening country of the Interior Plateau, the greater part of the ice was forced in this case to follow its length in the directions above indicated, instead of discharging laterally on both sides to the sea. A certain proportion of the ice, however, during the maximum phase of this great glacier, flowed through passes in the Coast Ranges, and uniting there with ice derived from the western slopes of these ranges, filled the great valley between Vancouver Island and the mainland, impinged upon the shores of the Queen Charlotte Islands, and still further north reached the ocean across the coast archipelago of the south-eastern coast-strip of Alaska.

Though at first in doubt as to the probable origin of the traces met with of this first and most general epoch of the glaciation of the Cordillera,[3] much additional information gained in later years convinced me that it clearly indicated the former existence of a great glacier-mass such as that here described.[4] Still more recent observations have proved the north-western movement of the northern part of the great glacier, and

---

[1] It is worth noting here that Mr. McConnell's carefully elaborated section through the Rocky Mountains on the line of the Bow and Kicking-Horse Rivers affords much ground for the belief that the central line of this range constituted a geological hinge even in Palæozoic times. This evidence consists chiefly in the difference in character met with in the formations of the western and eastern parts of the range.—'Annual Report Geol. Surv. Can.,' 1886, Part D.

[2] Such general movement probably affected only the central portion of the ice-mass by which the Cordillera was covered, and there is no reason to suppose that it was otherwise than sluggish.

[3] 'Quart. Journ. Geol. Soc.,' vol. xxxiv, p. 118.

[4] 'Quart. Journ. Geol. Soc.,' vol. xxxvii, p. 283.

after having thus ascertained its area, I ventured to designate it as the Cordilleran glacier.¹

The formation and movement of the Cordilleran glacier naturally resulted in the complete obliteration of the traces of the earlier and smaller glaciers which must have preceded it, as well as practically the complete removal of all unconsolidated pre-glacial gravels and sands. Of these last the only known remnants are the deep gravels of certain old streams in the more mountainous regions, which are in some places distinctly capped by boulder-clay, and have proved in the Cariboo District and elsewhere to be highly auriferous.

The limits in latitude above assigned to the great *névé* or gathering-ground and point of dispersion of the Cordilleran glacier are maximum limits, depending merely on the localities of observed traces of its action, and it is probable that detailed examination of the intermediate region will eventually admit of a much more precise localization of this area. It is further probable, arguing from the existing conditions of precipitation relatively to the Pacific, that the point of greatest accumulation was not more than 200 miles inland, while it is quite possible that it may not have been situated over 100 miles from the coast.² On the assumption that a slope of ten feet to a mile is necessary in order to produce motion in such a glacier,³ and taking into consideration the known length of the south-eastward moving portion of this great glacier-mass, its highest central part within the limits above assigned must have had an elevation of at least 7,000 feet⁴ above the mean elevation of the Interior Plateau, which would be equivalent to an elevation of about 10,000 feet above the present sea-level, or probably 11,000 feet above the sea-level of the time.

Nothing has yet been ascertained which might throw light on the question as to whether the great height of this *névé* was due simply to a local accumulation of ice, or whether its existence and the required degree of slope from it may in part be attributed to a greater amount of uplift which affected the particular region upon which it rested.

A brief statement of the known limits of the Cordilleran glacier may now be given. In 1878 the writer was able to state that if the general glaciation of the interior of British Columbia was attributable to the action of a confluent glacier (a point as to which he was at that time uncertain), the ice of its southern extremity must have poured southward through the gaps on the forty-ninth parallel.⁵ Beyond this parallel, which constitutes the international boundary, his investigations were not carried. This inference has since been confirmed, and the southern tongues or lobes of the Cordilleran glacier have in part been traced out by Prof. F. C. Chamberlin and Mr. Bailey Willis of the U. S. Geological Survey, the furthest southward extension being, as determined by Prof. Chamberlin, near the south end of Pend D'Oreille Lake in or about latitude 48°, 20'.⁶

---

¹ 'Geological Magazine,' Dec., III, vol. v., p. 348.
² In an article in the 'American Geologist,' vol. iv, p. 215, Mr. Warren Upham gives somewhat different limits for this area of dispersion, but as his statements are based upon my observations only, I am at a loss to understand his grounds for so doing.
³ Cf. Dana, 'American Journal Science,' III, vol. v, p. 205.
⁴ By assuming a minimum slope of one degree, in accordance with observations by Hopkins, this amount would be increased to about 31,000 feet.
⁵ 'Quart. Journ. Geol. Soc.,' vol. xxxiv, p. 119.
⁶ Cf. 'Bulletin No. 40 U. S. Geol. Survey,' 1887; "Seventh Annual Report U. S. Geol. Survey, 1888," p. 178.

On the coast, the extension of the Cordilleran glacier which occupied the wide valley between the highlands of Vancouver Island and those of the adjacent Coast Ranges, divided at a point a few miles north of Seymour Narrows, to form two broad glacier-streams flowing in opposite directions, or south-east and north-west respectively. These have been designated the Strait of Georgia and Queen Charlotte Sound Glaciers.[1] Of these streams of ice, the first-mentioned probably did not extend far beyond the south-eastern extremity of the island,[2] and it appears to be doubtful that it ever pushed southward so as to cover any considerable portion of the Puget Sound basin. The Queen Charlotte Sound glacier similarly appears to have terminated in the vicinity of the north point of Vancouver Island—Cape Commerell.[3] The extension of the border of the Cordilleran glacier to the Queen Charlotte Islands is believed to be indicated by certain striæ found near the northern extremity of these islands, the direction of which would show that the ice here advanced southward, in conformity with the main direction of the long fiords and channels between Observatory Inlet and Duke of Clarence Strait, which must have here been its principal feeders.[4] Still farther north, according to the observations of Prof. G. F. Wright and those of the writer, it is clear that the western border of the great glacier passed seaward across the coast archipelago of the southern part of Alaska.[5]

Having from an examination of the notes made by various arctic explorers arrived definitely at the conclusion that the great glacier-mass of the eastern part of the continent possessed a northward as well as a southward direction of motion from its main gathering-ground,[6] the writer was pleased to be able to avail himself of the opportunity afforded by the Yukon expedition to investigate the conditions of the northern part of the Cordilleran glacier. Evidence was there obtained of its northward or north-westward direction of movement, and this has since been confirmed and added to by observations in surrounding regions by Mr. R. G. McConnell of the Canadian Geological Survey (1888) and by Mr. I. C. Russell of the United States Geological Survey (1889).[7] On the Lewes and Pelly Rivers, branches of the great Yukon River, striated rock-surfaces, evidently due to the general Cordilleran glacier, were noted; in the case of the first-mentioned river as far north as latitude 61° 40' on the Pelly to latitude 62° 30', longitude 135° 45'. The observed bearings show a convergence of direction toward the low country about the confluence of these two rivers, near the site of old Fort Selkirk, and it is not improbable that the glaciers may have here reached to the vicinity of the sixty-third parallel on the 137th meridian. No traces of glaciation were observed by Mr. McConnell, still farther north, along the Porcupine River, nor by Mr. Russell further down the main valley of the Yukon, the appearances there being on the contrary those of a country which had long been subjected to subaerial decay, and

---

[1] 'Quart. Journ. Geol. Soc.,' vol. xxxvii, p. 278.
[2] This inference is drawn from the character of the ice action shown by the rocks in the vicinity of Victoria.—'Quart. Journ. Geol. Soc.,' vol. xxxiv, p. 95.
[3] 'Annual Report Geol. Surv. Can.,' 1886, p. 103 B.
[4] 'Quart. Journ. Geol. Soc.,' vol. xxxvii, p. 282 ; 'Report of Progress Geol. Surv. Can.,' 1878-79, p. 93 B.
[5] 'American Naturalist,' March, 1887; 'Geological Magazine,' Dec., III, vol. v, p. 348.
[6] 'Annual Report Geol. Surv. Can.,' 1886, p. 50 R.
[7] 'Bulletin Geol. Soc. Am.,' vol. i, pp. 540-188.
[8] 'Geological Magazine,' Dec., III, vol. v, p. 348; 'Annual Report Geol. Surv. Can.,' 1887-88, p. 40 B.

which had not been traversed either by glaciers or by floating ice capable of bearing erratics.[1]

Further illustration of the fact that the extreme north-western part of the continent remained a land surface, upon which no extensive glaciers were developed at any time during the Glacial epoch the time of maximum glaciation, is afforded by the note of Messrs. Dease and Simpson as to the entire absence of boulders along the Arctic coast westward from the estuary of the Mackenzie River.[2]

Granting that the north-western extremity of the Cordilleran glacier reached the furthest point above assigned to it, we find that its extension from the centre' gathering-ground (or from the approximate margin of this gathering-ground already given) was much shorter than that attained by the south-easterly flowing part, the approximate lengths being 350 and 600 miles respectively. This may be regarded as indicating either a greater relative elevation of this part of the continent to the north-westward, or a less copious supply of snow in that direction; the latter being the more probable supposition on account of the absence, which has just been referred to, of traces of glaciation in the extreme north-west.

The north-eastern margin of this great glacier is less easily defined, but it may (as a whole) be regarded as having been conterminous with the Rocky Mountains proper, against which it must have rested directly in some parts of its length, while in others the more or less isolated mountain-groups of the Gold Ranges doubtless constituted local gathering-grounds which contributed their quota to swell the main stream. It is certain that the great valley which separates the mountains of these ranges from the Rocky Mountains was throughout filled with ice, which had, like that of the main glacier, a south-eastward direction of movement in the corresponding part of its course (i.e., in that part of the valley which is now occupied by the upper parts of the Columbia and Kootanie Rivers). While it is possible that some part of this ice discharged laterally by the passes across the Rocky Mountains, this is rendered improbable by the absence of erratics derived from the Gold Ranges both in the Rocky Mountains and in the foot-hills to the eastward ? them.

No evidence has occurred to the writer such as to lead him to regard the boulder-clay he Cordilleran region as a *moraine profonde* of the Cordilleran glacier, and if the boulder-clay as a whole be not of this nature, but slight traces of any such bottom moraine are to be found. A hard stony material evidently of this character has been observed in some places near Victoria, wedged into crevices of the glaciated rocks or protected by their overhanging ledges; and in some of the deep V-shaped valleys further inland in this part of Vancouver Island, a very similar deposit sometimes forms a great part of the drift. In similar deep valleys and other sheltered low places in the Interior Plateau region, some much compacted boulder-clays to which this origin may with probability be assigned, and which may represent true till, also occur, but the general covering of boulder-clay appears to have a different and subsequent history.[3]

[1] 'Bulletin Geol. Soc. Am.,' vol. i, pp. 140, 543.
[2] "Narrative of Discoveries on the North Coast of America, 1836-39," p. 149.
[3] Mr. I. C. Russell in his "Notes on the Surface Geology of Alaska," which has already been referred to, seems to assume that I regard the boulder-clay seen along the Lewes, above Fort Selkirk, as a true glacier deposit. This assumption, however, does not precisely represent my view of its origin, which is alluded to on a later page.

The inferences above drawn respecting the existence and limits of the great Cordilleran glacier now rest upon a great body of facts of different kinds into the detailed consideration of which it is impossible to enter here. Our knowledge of the direction of motion and thickness of the glacier-mass, however, depends chiefly on observed instances of rock-striation or scoring met with on isolated high points or on the surface of broad plateau-like elevations. In the study of the interior region of British Columbia it became apparent at an early stage of the enquiry, that in addition to striation and shaping of rock-surfaces by glacial action referable to the various mountain-systems and diverging in all directions from these, traces of a much more general character and of older date also occur. The instances in which such evidence can be found under quite unequivocal circumstances are naturally rather infrequent. Some such were noted and described in my paper of 1878, others have since been published in later papers and reports, and some, met with during the season of 1889, have not yet been made public. In order, therefore, to present this important evidence in a concise manner, a number of the principal cases which have been discovered within the area of the south-eastward flowing portion of the Cordilleran glacier are tabulated below, the general order followed being from north-west to south-east. The approximate latitude and longitude of the localities is given, as most of these are not shown on the ordinary maps. Several of the mountains, indeed, have been named in the course of topographical and geological work still in progress, and are therefore not to be found on any published map. The names of these last are in parentheses.

LIST OF SOME PRINCIPAL INSTANCES OF STRIATION REFERABLE TO THE SOUTH-EASTWARD PORTION OF THE CORDILLERAN GLACIER, IN THE INTERIOR REGION OF BRITISH COLUMBIA.

PLACE.	Approx. Lat.	Approx. Long.	Height in feet above sea.	Direction of strie (true bearings).	REMARKS AND REFERENCE.
1. Summit of Tsa-whinz Mountain	53° 40′	123°	3,240	About S. 10° W.	An isolated point 800 feet above plateau between Chilacco and Fraser Rivers. Water-worn boulders and pebbles found. 'Rep. of Prog. G.S.C.,' 1875-76, p. 202; 'Q.J.G.S.,' vol. xxxiv, pp. 100, 104.
2. Summit of Sinter Knoll.	53°	125° 45′	3,550	S. 8° E.	An isolated point 250 feet above surrounding country. Wide lower plateau to the north. Erratics on summit. 'Rep. of Prog. G. S. C.,' 1876-77, p. 80; 'Q. J. G. S.,' vol. xxxiv, p. 101.
3. Spur of Tsi-tsutl Mountain	52° 40′	126° 10′	3,700	S. 37° W.	Ice here passed between Tsi-tsutl and inner border of Coast Ranges, crossing a 'col.' Lower country to the north. Direction somewhat affected by the local circumstances. 'Q. J. G. S.,' vol. xxxiv, p. 102.
4. Plateau north of Chilcotin River.	52°	122° 40′	3,650	S. 2° E.	The direction is transverse to the great gorge of the river. Several localities. 'Rep. of Prog., G.S.C.,' 1875-76, p. 201; 'Q.J.G.S.,' vol. xxxiv, p. 101.
5. High plateau between North Thompson and Bonaparte Rivers.	51° 02′	121° 13′	5,000	S. 35° E.	These are examples merely of the glaciation met with on prominent parts of this plateau. Erratics are strewn over all parts of the plateau. Of these examples all but No. 5 have already
6. " "	51° 08′	120° 40′	4,220	S. 30° E.	
7. " "	51° 05′	120° 35′	5,220	S. 34° E.	

Though peculiar in some places, in containing considerable masses of stratified clayey gravels, the boulder-clay of the Upper Yukon basin, where I have seen it, is often a typical boulder-clay of the character to be found over thousands of square miles in the interior of British Columbia and in the northern part of the Great Plains. *Cf.* 'Bulletin Geol. Soc. Am.,' vol. I, p. 143; 'Annual Report Geol. Surv. Can.,' 1887-88, pp. 126 B, 149 B.

List of some principal instances of Striation.—*Continued.*

PLACE.	Approx. Lat.	Approx. Long.	Height in feet above sea.	Direction of strie (true bearings.)	REMARKS AND REFERENCE.
8. High Plateau, etc......	51° 02′	120° 26′	5,450	S. 57° E.	been published, 'Geol. Mag.,' Dec., III, vol. vi, p. 352). As now given slight changes are made in some of the positions and heights of localities, which where formerly published were stated to be approximate, but have since been worked out with greater accuracy. No. 8 is the summit of (Skonti), a remarkable conical basaltic hill.
9. " "	50° 59′	120° 25′	5,840	S. 35° E.	
10. Tod Mountain.........	50° 56′	119° 55′	7,250	S. 44° E.	Culminating point between deep and wide valleys of North and South Thompson Rivers and Adams Lake. Lightly glaciated over summit. No travelled stones seen about actual summit. 'Geol. Mag.,' Dec., III, vol. vi, p. 351.
11. High plateau between Adams & Shuswap Lakes	51° 1′	119° 41′	6,100	S. 27° E.	A few small erratics seen, some on summit of point with height of 6,210 feet. This is also glaciated but the dir. in is indeterminable. 'Geo. Mag.,' Dec. III, vol vi, p. 352.
12. (Clear Mountains) between Hat Creek Valley and Fraser River........	50° 41′ 50° 36′	121° 42′ 121° 40′	7,070 7,040	None seen " "	Small erratics found everywhere strewn with local material up to the highest points of these mountains. Summits considerably shattered and weathered. Various parts of the range were examined, the positions and heights given are those of the northern and southern high points, (Chipooin) and (Blustry) Mountains.
13. Summit east of Paul's Peak, near Kamloops...	50° 42′	120° 14′	3,520	S. 51° E.	Travelled stones and considerable covering of drift material. Local topography explains divergence of glaciation to eastward.
14. Plateau 14 miles south of Kamloops..........	51° 31′	120° 24′	4,100	S. 31° E.	General direction from glaciated surfaces, the striæ weathered out.
15. (Cinder Mountain)....	50° 34′	121° 08′	5,070	S. 50° E.	Travelled stones on summit.
16. (Murray's Mountain)...	50° 31′	121° 33′	6,880	S. 10° E.	Neighbouring deep parallel valleys of Fraser and Thompson appear to have influenced direction. Boulders and stones of varied origin on summit.
17. (Spaist Mountain).....	50° 23′	121° 05′	5,780	S. 28° E.	Isolated high point on plateau north of Nicola River.
18. Zakwaski Mountain....	50° 00′	121° 16′	6,000	....	Isolated high point at head of Nicoamen River. Glaciated but direction uncertain, the trachytic rocks much broken up and weathered. Travelled stones 3 or 4 inches through found on summit.
19. High Point on ridge between last and Nicola River	50° 12′	121° 13′	5,030	S. 13° E.	Granite boulders strewn over this and neighbouring ridges, which are composed of volcanic rocks.
20. Iron Mountain.........	50° 03′	120° 45′	5,260	About S. 20° E.	Situated near confluence of Nicola and Coldwater Rivers. Glaciation heavy. Travelled stones on summit. 'Rep. of Prog. G. S. C.,' 1877-78, pp. 136 B, 146 B.; ' Q. J. G. S.,' vol. xxxvii, p. 272.
21. High point on plateau 20 miles south of Nicola Lake	49° 50′	120° 35′	4,380	S. 0° E. to S. 18° E.	Strie somewhat obscured by weathering.
22. Plateau near Chain Lake	49° 40′	120° 15′	4,075	S. 20° E. to S. 28° E.	Situated midway between Okanagan Lake and Similkameen River. ' Rep of Prog., G.S.C.,' 1877-78, p. 137 B.; ' Q.J.G.S.,' vol. xxxvii, p. 273.
23. (Loadstone Peak)......	49° 25′	120° 50′	6,370	S. 15° E.	Higest point in this vicinity to east of Coast Ranges.
24. Toad Mountain.......	49° 25′	117° 21′	6,090	S. 0° E. to S. 33° E.	This is in West Kootanie District and forms the watershed between Kootanie and Salmon Rivers. Glaciation light but distinct. No erratics observed. 'Ann. Rep. G.S.C.,' vol. iv, p. 40 B.

In further explanation of the above table, it may be pointed out that the less considerable height of the points in the northern part of the Interior Plateau upon which evidence of the general glaciation has been found, depends principally on the less elevated character of all that part of the plateau. Points equal in height to those enumerated to the southward scarcely occur, but this circumstance, with the less bold relief of this part of the plateau, in itself enables equally good evidence to be obtained at lower levels. It cannot be accepted as having any bearing on the probable thickness of ice during the maximum epoch of its accumulation in the northern and southern parts of the plateau respectively. The much greater number of instances drawn from the southern part of the region is due chiefly to the fact that portions of this part have now been closely and systematically examined, while the surveys carried out to the north (in 1875, 1876 and 1879) were of the nature of reconnaissances, and thus did not require nor even admit of the occupation and examination of all the high points. The same remark applies to the absence of observations at great heights in the area of the north-westerly flowing part of the Cordilleran glacier. It should also be remarked in this connection, that glacial striation on the summits of mountains such as most of those here cited, can generally be found only by close examination and search for unweathered rock-surfaces, and that many cases occur in which no certain indication of direction can be obtained even by such search.

Particular interest attaches to the observations on the elevated rough plateau between the North Thompson and Bonaparte Rivers (numbers 5–9), because of the fact that no considerable area of equal height occurs to the north-westward (from which direction the ice came) for a distance of about 350 miles. An inspection of the table will show that the main direction of motion of the part of the Cordilleran glacier represented by it was from north-west to south-east, along the Interior Plateau and parallel to the main mountain elevations of the Cordillera. Where least disturbed by local circumstances of the relief of the surface over which the glacier flowed, and by the occurrence of adjacent ranges which may have deflected the ice, the mean direction lies between S. 30° E. and S. 35° E. The general surface of the glacier must have stood at one time, in the southern part of the Interior Plateau, at a height somewhat exceeding 7,000 feet above the present sea-level, the thickness of the glacier-ice covering even the higher parts of the plateau here being thus at least 2,000 to 3,000 feet, while it attained a thickness of about 6,000 feet over the river-valleys and other main depressions of the surface. From the light character of the glaciation observed on most of the higher points, it is probable that the glacier did not much exceed the thickness here assigned, but that it preserved this thickness, together with its full width, to near the forty-ninth parallel, is indicated by localities 23 and 24 of the table, which are 170 miles distant from each other on an east and west line.

The erratics which occur on the summits of most of even the highest mountains on which glaciation has been found, as well as those scattered over high points where no striæ were detected, are as a rule to be classed as pebbles rather than as boulders. They are generally more or less rounded, but are occasionally striated. They are usually found sparingly dispersed among rocky debris, derived from the mountains themselves, and with little or no accompanying earthy drift. It is believed that these foreign stones were carried on the surface or within the mass of the upper parts of the Cordilleran glacier when at about its maximum, and that they were left stranded where now found as it declined.

Many observations might be cited to show that as the main glacier decreased in thickness and its front retreated, its direction of movement became more and more subsidiary to the local relief of the surface. This subject can, however, only be alluded to here. The general character of the change may be diagrammatically expressed as shown below, where the vertical lines represent the mountain ranges bordering the Interior Plateau, the arrows the direction of movement of ice at successive periods in its decline:—

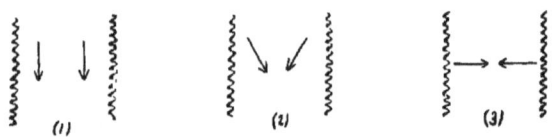

Evidence believed to be conclusive as to the lateral discharge of a portion of the ice through the valleys by which some rivers still traverse the entire width of the Coast Ranges, has been obtained on the Dean or Salmon River[1] (lat. 52° 50′), as well as in the larger inlets of the coast. The ice attained a thickness of at least 3,000 feet in the region where the Strait of Georgia and Queen Charlotte Sound glacier-streams diverged, between Vancouver Island and the mainland. Toward the extremity of the first-named glacier-stream, near Victoria, the ice must have had a thickness exceeding 600 feet.[2]

The directions in which the ice moved in the coast region have already been referred to.

Little can yet be said as to the thickness of the north-westward flowing part of the Cordilleran glacier, though the absence of heavy glaciation about the summit of the Chilkoot Pass (3,560 feet in height), by which the Coast Ranges are crossed in latitude 59° 45′, may be accepted as tending to show that the surface of the corresponding part of the main glacier could not have very much exceeded this height, a circumstance which would correspond with the inferior length of this portion of the glacier.

It has already been pointed out that the British Columbian portion of the length of the Cordillera, when the glacial epoch supervened, stood probably at least 900 feet higher than it now does. If this be admitted, as there appears to be every reason that it should, it will be found that the recession of the Cordilleran glacier and its accessory glacier-streams was contemporaneous with, if not brought about by, a movement of subsidence; for when the Strait of Georgia glacier had diminished only so far as to bare the glaciated rock-surfaces of the south-eastern extremity of Vancouver Island, these were at once covered by irregularly stratified deposits, comprising sands, clays, gravels and boulder-clay, in some of which marine shells are found.[3] Similar facts are observable further north in the Queen Charlotte Islands,[4] and it may thus be inferred that the land

---

[1] 'Quart. Journ. Geol. Soc.,' vol. xxxiv, p. 102.
[2] 'Quart. Journ. Geol. Soc.,' vol. xxxiv, p. 95.
[3] 'Quart. Journ. Geol. Soc.,' vol. xxxiv, pp. 90, 122.
[4] 'Report of Progress Geol. Surv. Can.,' 1878-79, p. 99 B.

had gone down at least 1,000 feet from the stage of its maximum elevation before any very considerable decrease in size of the great glaciers occurred. We are unable to follow this subsidence further in detail, as whatever traces it may have left along the coast must have been obliterated by subsequent action during the succeeding period of elevation which is referred to on a later page. We know only that in the Strait of Georgia and in Queen Charlotte Sound a considerable thickness of fine, regularly stratified silty material was laid down in a tranquil manner above the first boulder-clay and its associated deposits, under conditions implying that the land was at least from 100 to 200 feet lower than at present.[1] How much lower the land may have stood at this time we have as yet no evidence on the coast to show.

In previous publications I have classified the superficial deposits of the interior of British Columbia due to the Glacial period under the names *unmodified drift* and *modified drift*, the first-mentioned including the boulder-clay, the second embracing deposits of various kinds which have frequently in large part been formed by the rearranged materials of the boulder-clay. Though the division thus made is not in all cases perfectly definite, it is warrantable from a general standpoint, and convenient for purposes of description.

The boulder-clay though not differing in any obvious manner from that of the Great Plains, or indeed in any important respect from that generally found in different parts of the Northern Hemisphere, presents here some minor peculiarities. It consists generally of a paste of hard, sandy clay, containing usually a very considerable proportion of fine sandy material, through which stones of all sizes are irregularly scattered. Its colour as a rule varies from light brown to pale yellowish or grayish brown, but in freshly exposed sections is sometimes bluish-gray. It has usually a more earthy appearance than that of the eastern part of Canada, and very often over extensive regions forms the soil in which the trees are rooted, without the intervention of any modified drift. An unusually large number (including in fact much the larger part of the whole) of the stones and boulders are well rounded and water-worn, but a variable proportion showing distinct and sometimes heavy glacial striation or polishing is constantly present.

As later and more extended observations have served only to bear out the description of its mode of distribution given in a former publication, this may be quoted as originally written:—" Over considerable areas this material is concealed beneath the accumulations which form terraces and low-level flats, in relation to former lake and river-valleys. There is a remarkable uniformity in these boulder-clays in every locality in which I have examined them. In many places they form low rolling and broken hills between the river-troughs above the level of the higher terraces. In this case they appear sometimes to be spread in a comparatively thin layer over a rocky substratum; while in others they are of great depth, and by the irregularity of their arrangement have themselves produced many of the minor features of the surface. They frequently show a tendency to form more or less well-defined high-level plateaux, and are spread almost universally over the elevated basaltic region of the interior, in most places so uniformly, notwithstanding minor irregularities, as to allow the underlying rock to be very seldom seen."[2]

---

[1] 'Annual Report Geol. Surv. Can.,' 1886, p. 105 B.
[2] See 'Quart. Journ. Geol. Soc.,' vol. xxxiv, p. 103.

These remarks refer particularly to the Interior Plateau region, in the southern part of which later observations show that considerable differences exist as between various areas in the average depth of the boulder-clay deposit. In the Upper Yukon basin, again, it has been found that the boulder-clay presents some additional peculiarities, particularly in very often containing intercalations of clayey or earthy gravels which are evidently contemporaneous in date and into which the true boulder-clay is found to pass. This is, however, again referred to on a later page.

Throughout the Interior Plateau the upward limit of the boulder-clay is found at a height somewhat greater than 5,000 feet above the present sea-level, and corresponding in this respect with the highest level of well-marked terraces; the higher terraces in fact generally consisting of material identical in character with that of the general covering of boulder-clay, or so closely alike as to be indistinguishable from it. Though, as previously noted, travelled stones occur on much higher points, no boulder-clay, and very little fine drift material of any kind, has been found above the highest terrace-level referred to.

These highest terraces may be stated to have an average elevation of about 5,290 feet. Lower terraces, ranging between 5,000 and ⸺ feet have also been found in a certain number of widely separated localities, but the principal development of terraces is found below 3,800 or 3,500 feet, and especially when levels of 3,000 feet and under are reached. Below a height of about 3,000 feet the whole Interior Plateau region may be described as terraced, and although along the various river-valleys many terraces occur which have evidently been produced by the streams themselves while cutting down through the drift deposits, which at an earlier stage had filled these valleys, these need not be considered here, the point to which it is wished to draw attention being the existence of terraces requiring for their explanation a general flooding of the country. Such terraces are found to be not confined to the immediate valleys of the rivers, but to occur in different situations along the higher slopes, and to fringe at similar elevations the various irregularities of the plateaux.

The existence of that which has been referred to as the highest terrace-level was first ascertained in 1876, on the upper slopes of the Il-ga-chuz Mountain, in latitude 52° 45′. This terrace or beach-line has already been fully described,¹ and it need here be mentioned only that its elevation is 5,270 feet. The circumstances of observation at this place appear to be perfectly unexceptionable, though the terrace found here has remained for a long time an isolated instance.

During the progress of geological work in the southern part of the Interior Plateau in 1888 and 1889, additional information has, however, been obtained respecting this or other similar very high terraces. No publication of the results of this work having yet occurred, it will be necessary to refer briefly to the observed facts.

Tod Mountain has already been mentioned in connection with glacial striation, and its isolated position has been noted (see p. 32). A narrow and rough, but fairly well defined terrace occurs on its south side at a height ascertained to be 5,116 feet.

Nearly in the same latitude with the last, but twenty-eight miles further west, on the edge of the plateau to the south of the lake in which the Tranquille River rises, a terrace occurs at a height of 5,340 feet. This is perfectly distinct and somewhat extensive, and was seen from a distance to be repeated on the slope of the plateau some miles to the north-

---

¹ 'Quart. Journ. Geol. Soc.,' vol. xxxiv, p. 107 ; ' Report of Progress Geol. Surv. Can.,' 1876-77, p. 36.

ward beyond the lake. Where examined, this terrace is found to be composed of boulder-clay or identical material.

On Murray's Mountain, situated about fifty miles distant from the last, on a south-east bearing, in latitude 50° 31′, and which forms a summit on the mountainous ridge between the Fraser and Thompson Rivers, a terrace at a height of 5,376 feet occurs. It is situated in a depression on the southern side of the mountain, and though small and somewhat irregular is perfectly distinct.

Forty-six miles north of the last-mentioned locality, on the west side of the Marble Mountain range, in latitude 51° 8′, is a distinct terrace at an elevation of between 5,260 and 5,360 feet.

The clearest and most unequivocal evidence of the existence of a water-line at this great elevation was, however, found in October last on the eastern or opposite slope of the Marble Mountains. This looks out upon the sea-like expanse of the basaltic Green Timber Plateau, which has an elevation of from 3,800 to 4,000 feet. The entire eastern base of the range, for a length of fifteen miles, is heaped with drift deposits which are more or less distinctly terraced all along, the appearance presented from a distance being much like that of the Ilgachuz Mountains when similarly viewed.¹ When closely examined this accumulation of material is found to consist in part of moraine mounds and ridges, many of which have been more or less completely modified in form by water action, in part of sandy and gravelly terraced flats. The highest of these terraces, near the southern end of the mountains, range from 5,300 to 5,500 feet. The highest observed terrace toward the northern extremity of the mountains (twelve miles distant from the last) had an elevation of 5,100 feet.²

Mention may next be made of a few notable examples of lower, but still very high terraces met with in the interior of British Columbia or in that part of the Cordilleran region to the north of the province.

In the southern part of the Interior Plateau, the following instances have been noted:—Between heads of Guichon and Three-Mile Creeks, distinct terrace at 4,150 feet. Valley of Ray Creek, a tributary of Guichon Creek, distinct terrace at 4,350 feet. Along west side of lower part of Guichon Creek valley, terraces at about 4,000 feet. On plateau north of lower part of Nicola River, highest terraces at 4,396 feet. In valley of Prospect Creek, a tributary of the Nicola from the south, terrace at 4,660 feet. On hills to east of lower part of Hat Creek, terrace at 4,300 feet. On Green Timber Plateau, north of Clinton, sandy terrace-flats at heights of 3,900 to 4,150 feet. Further north, near Tatlayoco Lake, well-marked terraces occur at a height of about 4,250 feet.³ Between Skeena Forks and Babine Lake, in latitude 55° 20′, a wide terrace-flat was found at an elevation of about 4,300 feet, and other similar terraces, one of which was estimated at about 4,900 feet, were seen at a distance.⁴ On Dease River the highest observed terrace is at an elevation of about 4,600

---

¹ See illustration 'Quart. Journ. Geol. Soc.,' vol. xxxiv, p. 100.
² The only cases noted which have the appearance of indicating terraces at higher levels than those here described occur upon Tod Mountain. Two very small terrace-like flats, lodged in depressions scanning the mountain side, at heights of 5,050 and 5,720 feet respectively, were here found. These, however, appeared to be composed of fine earthy material, and might have been produced by the washing down of debris against masses of snow, which may at some time have occupied the depression referred to. Their existence cannot be accepted as possessing any significance in connection with the general questions here discussed.
³ 'Quart. Journ. Geol. Soc.,' vol. xxxiv, p. 108.
⁴ 'Quart. Journ. Geol. Soc.,' vol. xxxvii, p. 276 ; 'Report of Progress Geol. Surv. Can.,' 1879-80, p. 137 B.

feet.[1] In latitude 61° 40', on the watershed between the Liard and Yukon River systems, terraces occur to a height of 4,300 feet, or somewhat higher, and the summit of an isolated mountain of the height just mentioned was found to be strewn with rolled stones of varied origin, the circumstances being such as to show that water must have stood at one time about 1,000 feet above this part of the Pacific-Arctic watershed:[2]

The above instances all refer to the Interior Plateau of British Columbia, or to the region similarly situated as respects the marginal ranges of the Cordillera, to the north, and it will be noticed that (with the exception of Tatlayoco Lake) none of the higher terraces have been found on the eastern flanks of the Coast Ranges, which, it may be presumed, were covered by glacier-ice at the time of formation of these terraces. It may further be stated that, as a whole, the higher terraces referred to in the foregoing paragraphs are comparatively rare, and that many of them show the effect of considerable denudation, while below levels of 3,800 or 3,500 feet terraces are extremely abundant almost everywhere, and are very frequently wide and in an excellent state of preservation. The heights of the terraces above noted have been barometrically determined within small limits of error, the influence of the weather at the time of observation having been eliminated by the use of station barometers. An investigation of the lower terraces, carried out by more refined means of measurement, might produce interesting results, but for the purposes of this sketch it will be unnecessary to enter into the great mass of observations which has been accumulated respecting the heights of these in different localities. Before leaving this branch of the subject I would, however, mention the fact that water-rounded stones occur on the slopes of the mountains at the summit of the Pine Pass across the Rocky Mountains (lat. 55 20'), several hundred feet above the actual summit, and that an apparent terrace was noted at 300 to 500 feet above the same summit, or 3,300 to 3,500 feet above the sea-level. Allusion may also be made to the terraces at great heights met with on the eastern slopes of the Rocky Mountains, which are again referred to on a subsequent page.

In the paper on the "Superficial Geology of British Columbia," published in the 'Quarterly Journal of the Geological Society' for 1878. to which reference has frequently been made, it is stated that I had not up to that time met with any distinct indications of moraines which might be referred to a great Cordilleran glacier.[3] Observations made during the past two years, however, have supplied some evidence of such moraines, which, though not of great dimensions, appear to be definitely referable to the period of retreat of the southern extension of the great glacier. These are distinct from a much larger class of moraines produced at a later date, and due to local glaciers moving from the several mountain ranges. Moraines referred to the Cordilleran glacier are found on the plateau near the sources of Otter River (latitude 49° 45'), and again fifteen miles further north near the head of McDonald River. They produce in some places a lumpy irregular country, but are also found aligned in parallel series, running north and south, and in this case evidently representing lateral moraines produced at the edges of a residual tongue or lobe of the great glacier, which was gradually being reduced. Where arranged in tiers, these lateral moraines are sometimes separated by narrow V-shaped valleys only

---

[1] 'Annual Report Geol. Surv. Can.,' 1887-88, p. 96 B.
[2] 'Annual Report Geol. Surv. Can.,' 1887-88, p. 119 D.
[3] 'Quart. Journ. Geol. Soc.,' vol. xxxiv, p. 119.

somewhat resembling those formerly described on the Nechacco.¹ In height, the ridges seldom reach one hundred feet. Smaller transverse moraines, sometimes hollow to the north, were also observed. On the Green Timber Plateau, north of Clinton, numerous similar moraine ridges occur and may sometimes be followed continuously for a mile or more. In the region still further north I am now inclined to regard the ridges which characterize the low country (fifteen miles in width) between Il-ga-chuz and Tsi-tsutl Mountains as moraines produced during the retreat of the great glacier.² All these moraines are formed of material resembling that of the boulder-clay, but containing apparently a somewhat larger proportion of water-rounded stones.

It is in endeavouring to follow the course of events subsequent to the period of culmination of the Cordilleran glacier that our investigation becomes more than elsewhere beset with difficulties. These arise in great part from the varied and bold orographic features of the Cordillera, which render it necessary to have in view the local circumstances of each observed fact in a manner not found so essential in a less broken country. There is necessarily much difficulty in assigning a relative order to the various phenomena, and it is especially difficult in some cases to separate those due to the epoch of the Cordilleran glacier from those brought about by a later advance of glaciers from the various mountain ranges, which without doubt occurred. It has for this reason been considered necessary to review briefly, in foregoing pages, the phenomena believed to be clearly connected with the first and most severe epoch of glaciation before attempting any explanation of the mode in which these have been brought about. It is particularly in a complex and vast region like that here treated of that the accumulation of a great number of observations, some of which appear to be incompatible one with another, embarrasses any process of generalization. With the equipment of a limited number of observations only, derived from some single portion of the region, no great difficulty might be found in including these in some logically consistent scheme which might or might not eventually prove to be correct. While therefore some general account of the observed facts has been given, it is impossible to extend this so as to include the description and discussion of the whole number, and the writer, in having to bear in mind a mass of detail with which this paper cannot be encumbered, labours under some disadvantage in drawing conclusions which the reader may feel able to accept. from that part of the evidence which has been brought to his notice. This difficulty is added to by the circumstance that it appears to be desirable to study the glaciation of the Cordilleran region on its own merits, divesting the mind as far as possible from hypotheses advocated or received for the eastern half of the continent, with which there is no *a priori* ground for believing that the events of glaciation of the Cordillera were identical, but rather some basis for the belief that they were complementary.

In previous publications I have suggested two modes by which the production and arrangement of the boulder-clay and other superficial deposits may be explained,³ but subsequent and more extended observations appear to show that neither of these is fully satisfactory. I will here first mention these suggestions, and then point out the main facts which appear now to stand against them:—

---

¹ See illustration 'Quart. Journ. Geol. Soc.,' vol. xxxiv, p. 109.
² 'Quart. Journ. Geol. Soc.,' vol. xxxiv, p. 113; 'Report of Progress Geol. Surv. Can.,' 1876-77, p. 36.
³ 'Quart. Journ. Geol. Soc.,' vol. xxxiv, p. 119; vol. xxxvii, p. 283.

(1) The general subsidence of the Cordilleran region, which we found reason to believe was in progress at the time at which the glaciers began to diminish, may have continued, till, at a later date, when the main Cordilleran glacier began to retreat from the Interior Plateau, the sea stood at the level of the highest observed terraces, or at 5,290 feet above its present level. In this case the boulder-clay might have been formed along the decaying front of the glacier, in part directly as morainic material, in part in a secondary manner from the droppings of icebergs floating upon the sea. The terraces, under this hypothesis, have been formed in order, from highest to lowest, as the land again rose to its present level.

This hypothesis has the merit of simplicity and would account reasonably well for most of the phenomena. As against it, however, it may be argued that it is difficult to understand how the moraine-ridges previously described succeeded in maintaining their identity under such a depth of debris-bearing sea. The absence, so far as observed, of any marine shells in the drift deposits, is also adverse to the acceptance of this hypothesis; for the sea must have had very free access to the Interior Plateau during the maximum of such a subsidence. Still further, the amount of subsidence required may be considered as so great as to be almost, if not quite unparalleled elsewhere, when the short time within which it must have been accomplished and reversed is taken into account.¹

(2) It is conceivable that, when the southern part of the Cordilleran glacier had abandoned the Interior Plateau, local glaciers, developed in the Coast Ranges, the Rocky Mountains and on the mountainous barrier to the southward near the forty-ninth parallel, stopped the various passes and valleys so completely as to give rise to the formation of a great glacial lake, the northern limit of which was the retreating front of the main glacier. The surface of this lake may have stood at the maximum elevation above noted, and debris-bearing ice might then have floated freely on it in various directions. As the glaciers by which it was held in became reduced, such a great lake might have been gradually drained, the series of terraces being formed as under the preceding hypothesis, but without necessarily implying that the sea stood any higher relatively to the land than at present.

The arguments against this hypothesis appear to me, however, to be insuperable. It is true that, with the exception of the Crow Nest Pass (about 4,830 feet) and Yellow Head Pass (3,700 feet), the gaps in the Rocky Mountains to the south of the Pine and Peace River Passes now stand as high or higher than the most elevated terraces of the Interior Plateau, but these two passes, at least, to the eastward must have been stopped, in addition to all the low valleys which traverse the Coast Ranges. The greatest difficulty occurs, however, to the south, where in particular the wide Okanagan Valley, with a height of only 880 feet on the forty-ninth parallel, would require to be closed by some portion of a glacier which no high mountain ranges are at hand to provide. It is thus apparent that this hypothesis requires an almost inconceivable co-ordination of glacier-dams, besides which there is really no valid evidence to show that glaciers are capable of holding in

---

¹ It should be borne in mind, however, that in times geologically recent, very great changes in relative level of land and sea can be shown to have taken place, and that it is therefore unsafe on any a priori grounds to exclude such great local changes from consideration. Mr. Upham, collecting a number of authenticated cases, which need not here be enumerated, notes post-Pliocene elevation or depression of 1,000, 1,000, 2,000, 2,800 and 3,000 feet. Wright's "Ice Age In America," p. 582. See also a paper by the Duke of Argyll, 'Scottish Geographical Magazine,' vol. vi, p. 177.

any very great depth of water, and much reason to believe that they are incompetent to do so.[1] In addition to this, such an hypothesis is even less adapted to account for the body of water beneath which a great part of the Yukon district must have been submerged when the northern end of the great glacier left it. Still further, it leaves us without any apparent cause for the decay of the Cordilleran glacier at this particular period, the subsequent partial resumption of glacial conditions, and their final disappearance; all which events would require to be attributed to some general climatic or cosmic series of changes.

A third hypothesis which I now venture to suggest combines some of the features of the two first and appears to me to include all the observed facts better than either of these, and to form indeed a reasonably satisfactory explanation of the phenomena with which we have to deal :—It may be supposed that as a consequence of, or correlatively with, the gradual subsidence of the northern part of the Cordillera, the supply of snow producing and maintaining the great glacier became smaller, and that more of the winter increment was melted away during the summers, till at length the glacier itself became nearly stationary. Its decay, still continuing. resulted eventually in the formation of *englacial lakes*.[2] These might be expected to occur in the central parts of the Interior Plateau at a considerable distance from any of the bordering ranges, which still doubtless continued to contribute a certain quantity of ice to the mass. This central part of the Interior Plateau is, besides, that characterized by least precipitation at the present time, and consequently that in which the want of a continued re-supply of ice from the main *névé* would first become apparent. Such lakes, it may further be inferred, would originate in the first instance near projecting mountains or minor mountain ranges of the plateau region, all which conditions are in accordance with the mode of occurrence of the observed highest terraces. The terraces themselves may have been somewhat rapidly built, by the washing down to the water-line of material which had previously accumulated on the higher slopes of the projecting points, as well as from deposits borne by floating ice. The similarity in elevation of these highest terraces would appear to imply that the bodies of water thus formed within the area of the glacier were in more or less complete connection, and if all the observed terraces at about the level of 5,290 feet in the southern part of the Interior Plateau may be supposed to have owed their origin to a single lake, this must have had a length from east to west of about eighty-eight miles, with a north and south width of about fifty miles. This does not include the terrace or beach on Il-ga-chutz Mountain, much further north; and though it is possible that this also may have been formed in a part of the same lake, it seems more likely that a separate lake opened here, the height of which may have been in relation to the elevation of the general surface of the glacier at the time. Other terraces which have been noted at heights down to 4,000 feet or thereabouts, may be attributed to later stages of the same or similar lakes produced in the central parts of the glacier, and it may be that some or all of these lakes were comparatively shallow, being floored as well as walled around by the mass of the glacier.

At what particular stage in the decay of the great glacier these englacial lakes were

---

[1] Though it appears to be frequently taken for granted that glacier-ice is capable of holding in great inland seas, there is really little warrant in nature for such a belief, such miniature instances as Merjelen See being scarcely cases in point.

[2] This term is employed as a convenient one for such lakes developed on the surface of a great glacier as those found by Nordenskjöld on the inland ice of Greenland.

finally drained it is impossible to say, but it appears probable that this must have happened before the glacier ceased to cover the greater part of the Interior Plateau. It may further be supposed that the general subsidence of the Cordillera before alluded to progressed *pari passu* with the decay of the Cordilleran glacier, and there is some evidence (which, however, cannot be given here in detail), afforded by terraces formed contemporaneously with moraines, to show that the subsidence had carried the land down to a stage about 3,000 feet below the present sea-level, while tongues of the great glacier still extended as far south as latitude 49° 40′. Further evidence of the same kind favours the belief that by the time the end of the glacier had retreated about one hundred miles to the north, or to latitude 51° 30′, the subsidence had progressed to about 3,800 or possibly to 4,000 feet, which was probably about its maximum. The great glacier must have retreated rapidly toward the close of the first period of glaciation, and have become reduced to systems of small local glaciers in the mountain regions, unless indeed the central portion of the *nevé* may have retained a confluent character till after the second period of glaciation.

Along the retreating front of the glacier, and subject to a certain amount of rearrangement by the water which washed its base, the boulder-clay appears to have been laid down, and is, as before stated, indistinguishable in general character from the earlier and higher deposits of the same kind attributed to the euglacial lakes. The land can not have long remained at the low level which has been above assigned to it, the movement in subsidence being immediately followed by one in progressive elevation, during which all the more obvious and well-preserved terraces of the Interior Plateau and other parts of the entire southern portion of British Columbia were formed.

Before, however, following this presumed re-elevation further, we may glance for a moment at the condition of the northern part of the Cordilleran glacier during the supposed period of greatest subsidence. It has already been mentioned that the boulder-clays of the Upper Yukon basin present certain peculiarities. While in many places along the Upper Pelly and in most instances along the Lewes River the boulder-clay is of a typical character, it is often on the former river and sometimes on the latter represented by earthy or clayey, gray or brownish, stratified gravel-beds. These are found to pass horizontally into true boulder-clay, while in other instances they are interbedded with rude layers of boulder-clay, or form the lower or upper members of sections showing a considerable thickness of boulder-clay. The evidence I believe to be conclusive that they constitute with the boulder-clay a single formation, which represents the first deposit of the retreating northern extension of the Cordilleran glacier.[1] The stones both of the stratified gravels and the more typical boulder-clays are generally in this region well rounded, and glaciated stones or boulders are comparatively scarce. The stratified earthy gravels are moreover most abundant in the higher parts of the country traversed by the Upper Pelly River, at levels between 2,500 and 3,000 feet.

The character of the deposits representing the boulder-clay period in this northern region are taken to indicate that the total amount of subsidence was there less than to the south, that the material dropped along the front of the retreating ice-foot fell into shallower water, and that in conformity with this circumstance and the less perfectly enclosed character of this northern region and more diffuse arrangement of the mountain

---

[1] 'Annual Report Geol. Surv. Can.,' 1887-88, pp. 119 B, 126 B.

ranges there, it was subjected to a greater degree to current action.¹ Still further to the north and north-west, reasons have already been given for the belief that the land was not generally submerged at this time. It appears to be quite probable, however, that a wide strait was opened along the Yukon Valley to Behring Sea.²

It should be noted, before going further, that the fact that marine shells have not been found in the deposits which are supposed to have been laid down along the retreating front of the great glacier in water in communication with the sea and governed in elevation by it, may at first sight appear to constitute an argument equally weighty against the hypothesis above stated as against that of a general submergence to an extent of over 5,000 feet. On consideration, however, it will be found that a submergence of the Interior Plateau to a depth not exceeding 3,800 or 4,000 feet would still leave that region well enclosed by mountain ranges, and bearing in mind the quantity of fresh water which must have been poured into it, the necessary coldness of its water, and the great amount of earthy matter which was being discharged into this water, and comparing these conditions with their nearest known analogues, we are led to suppose that marine life must have been very scarce if at all able to maintain itself. These arguments, however, do not apply with equal force to the region of the Yukon basin, and here, if anywhere, we may eventually hope to find evidence of such life.

When the re-elevation of the land succeeding this great subsidence had progressed to a certain extent, the increased height of the mountain ranges of the Cordillera, in combination with the more general causes of glaciation of the northern hemisphere, became such as to lead to the increase and re-advance of their local glaciers, and probably also to a resumption of southward and northward movements from the central gathering-ground of the great Cordilleran glacier. Evidence of this second advance of ice is found particularly in the Interior Plateau region, where the growing glaciers have pushed out from the mountains along the various river-valleys, ploughing up previously formed terraces and gravel deposits and piling them into moraine ridges.³

On the coast, evidence of like bearing is afforded by the upper boulder-clay of the northern part of the Strait of Georgia, which may find its representative in the southern part of the same strait in the scattered large erratics, which are everywhere abundant there, in or overlying the highest layers of the drift deposits.⁴

---

¹ In his notes on the "Surface Geology of Alaska " 'Bulletin Geol. Soc. Am.,' vol. i, p. 143, Mr. I. C. Russell alludes to the boulder-clay seen by him on the Lewes as doubtfully representing true boulder-clay. I may say, however, that I feel assured that to anyone acquainted with the boulder-clay of British Columbia and the Great Plains such no doubt would be admissible. I may add that Mr. McConnell, who is perfectly familiar with these northern boulder-clays, coincides in my reference of those of the Lewes.

² Some evidence of this is indeed suggested by Mr. I. C. Russell (Op. Cit. p. 139), who writes :—" It may be well in this connection to direct attention to certain obscure indications of terraces or sea-cliffs, at an elevation of fifteen hundred or two thousand feet, on a number of the mountains near the Yukon, below Nulatto. None of these mountains have been closely examined, and it is impossible to state whether the indefinite lines which may indicate terraces are horizontal or whether they coincide in elevation. It is not safe to assume that they are terraces, as it is possible that they may indicate lines of structure or be due to land slides. The mountains are so situated that they could not have retained a lake, and if water lines exist on them their origin must be looked for in a submergence of the land."

³ Cf. 'Quart. Journ. Geol. Soc.,' vol. xxxiv, p. 113. It should be stated that subsequent investigations have led me to doubt the validity of the evidence in the cases first mentioned in the publication here referred to, but have at the same time added many additional facts, which cannot here be detailed, in confirmation of the general proposition.

⁴ 'Annual Report Geol. Surv. Can.,' 1886, p. 105 B.

The extent to which the land of the Cordilleran region may have been elevated at this time is not known. Judging alone from the size of the glaciers produced, it must have been much less than during the preceding maximum time of glaciation. In this case the coast region may still have remained submerged and the upper boulder-clay of that region may have been produced as the ice finally retreated. If, however, some general change toward amelioration of climate was in progress, an amount of elevation equal to that of the first great period of glaciation may have been required to produce the smaller glaciers of the second epoch. Some indication of an elevation equal to or greater in amount than that now held by the land is afforded by the notable absence of boulder-clay in parts of the larger valleys to the east of the Coast Ranges.[1] By such an elevation the removal of this deposit in these places may be accounted for, while the deposition of the upper boulder-clay of the littoral may have occurred during the partial subsidence which inaugurated the deposition of the white silt formation of the interior, about to be referred to.

To the latter part of this second and less intense, though possibly more protracted, epoch of glaciation we may with confidence assign the origin of the deposits which I have designated in previous reports and papers as the "*white silts.*" These were first examined in some detail in the basins of the Nechacco and Chilacco Rivers, in the northern part of the Interior Plateau, but subsequent and wider observations appear to show that these silts possess a greater significance than originally supposed, and that they serve to mark out pretty definitely a synchronous period of stability which extended to the entire northern part of the Cordillera, and which succeeded a partial subsidence of the lately re-elevated land.

In the regions characterized by them, which are in almost all cases at a less elevation than 2,500 feet,[2] these white silts very often rest directly upon the boulder-clay. They are generally fine and uniform in texture and are usually well bedded in perfectly horizontal layers of an inch to two or three inches each in thickness. Where occasional sandy or gravelly layers are intercalated, these are attributable to local causes, being most frequently found opposite the mouths of valleys down which streams have flowed. In some places, and particularly in certain sections along the Upper Pelly River, the layers have suffered crumpling or disturbance, apparently from the action of grounding ice, to which the rare occurrence of stones and boulders of considerable size may also be attributed. The silts have evidently been laid down, as a rule, in tranquil water of considerable depth, and their material has as obviously been supplied by streams or rivers discharging from glaciers not far removed. In physical characters the silts resemble the deposits of the Red River Valley, though usually in the Cordilleran region paler in colour, and seldom so clayey as some parts of those of the Red River. They differ from loess chiefly in their well-bedded character. It is believed that the general correspondence in elevation of the various and more or less separated bodies of water in which this white silt formation was formed, in itself constitutes a strong argument in favour of the hypothesis that these bodies of water were in direct communication with the sea and were governed in their

[1] The removal, more or less complete, of boulder-clay from these valleys can not be referred to the time of post-glacial elevation spoken of further on, as the white silt formation is found well developed in these valleys.
[2] In this statement I omit the consideration of a few instances of the local occurrence of similar silty material in the southern part of the Interior Plateau, which are not connected with the main development of silts. These serve, however, to connect this main period with an earlier time of greater flooding of the Interior Plateau.

level by that which it held at the time. No traces of morainic or other barriers have been found in any case sufficient to account for the damming back of water at the requisite level, nor do the local circumstances admit the supposition that such water was held in by glacier-dams. Had the silts been formed merely in lakes produced in one or other of the modes last mentioned, they might be expected to occur, in a region with such strongly marked features as that of the Cordillera, at a variety of very different levels, in correspondence with circumstances varying in each particular case. The length of the period required for the deposition of a great thickness of these fine beds also affords reason for belief in the long epoch of stable conditions, and to some extent justifies the presumption of the proximate contemporaneity of such a stable period. On following the silt formation toward the various mountain ranges and sources of local glaciers, it is almost invariably found to be cut off somewhat abruptly before the mere increase in elevation would account for its disappearance. This circumstance may with very little doubt be attributed to the fact that during the deposition of the silts these upper parts of valleys were occupied by the still considerable local glaciers of the second period, which were engaged in producing the material of the silts. The evidence is, further, conclusive that these glaciers in the end retreated pretty rapidly, leaving, in many cases, the long trough-like valleys which they had occupied almost entirely free from debris or detrital matter of any kind, and ready to become the beds of fiord-like lakes.[1]

Assuming the general contemporaneity of the several developments of the silt formation, and in view of the facts last stated, we are furnished with the means of ascertaining the extent of the glaciers, and ice-covered regions during a phase of the second glaciation, which, however, does not represent the maximum of that period, but a prolonged pause in its decline. In support of this belief, it may be stated that the indications thus arrived at agree well with the relative importance which we would on other grounds be justified in assigning to the glaciers of the different parts of the Cordillera.

For the purpose of illustrating the character of the evidence afforded by the silt formation, the principal areas of its occurrence may now be noted, proceeding from south-east to north-west. It is obvious, that in respect to the level at which the water stood at the time the silts were laid down, the important elevations are those marking the maximum heights attained by the deposits in question, in their larger areas, as the lower parts of the deposits may have been produced in water of very considerable depth.

In the great Columbia-Kootanie valley, lying between the Rocky Mountains and Gold Ranges and opening widely to the southward, the silts are well represented, with a thickness in some cases of fifty to one hundred feet as shown in terraces since cut through them, and a maximum observed elevation, near Upper Columbia Lakes, of about 2,700 feet. They are seen at lower levels down to 2,200 feet or perhaps less.[2]

In the southern part of the Interior Plateau region, in consequence of the considerable height of the mean level of the country, the white silt formation is usually confined to

---

[1] 'Report of Progress Geol. Surv. Can.,' 1877-78, p. 153 B; 'Quart. Journ. Geol. Soc.,' vol. xxxvii, p. 275.

[2] Silty deposits similar to those of the main valley are found in limited quantity in the Rocky Mountains to the east of it, in the narrow tributary valleys of the Kicking-Horse and Wigwam Rivers to a height of about 3,500 feet. I am now inclined to refer these to an earlier period than that of the main area of white silts, possibly even to the closing period of the first maximum of glaciation. Their existence in these mountain-valleys may, however, be accepted, in any case, as evidence that the local glaciers developed on the Rocky Mountains proper during the second period of glaciation were not of great size.—'Annual Report Geol. Surv. Can.,' 1885, p. 30 B.

the various trough-like valleys. It may be seen from the line of the Canadian Pacific Railway in characteristic development, forming frayed terraces along the South Thompson valley for a number of miles east of Kamloops. It is found also along Okanagan Lake, but in the southern continuation of the Okanagan Valley, near the forty-ninth parallel, is chiefly represented by fine sands. Along the lower part of the Similkameen it has been observed in patches. On the Nicola it is often well displayed. It extends down the main River Thompson to about the mouth of the Nicola, and appears again characteristically developed along the Fraser to the west of Clinton. It stretches far up the North Thompson, and runs back along the valleys of the main tributaries of this stream as well as of those previously mentioned. The silt formation is often along these valleys a striking feature, and is shown in terraces from 100 to 200 feet or more in height.

It is worthy of remark, that on the lower part of the Okanagan Valley, as well as on the lower parts of the main Thompson and Fraser, the silt is reduced in quantity and replaced by coarser arenaceous deposits, a fact tending to show the existence of rather strong current-action in these main outlets of the plateau region. The silt formation is entirely absent from the upper portions of the valleys occupied by Adams Lake, Shuswap Lake with its various arms, the Arrow Lakes, and the northern part at least of Kootanie Lake.[1] Mabel and Sugar Lakes, lying between Shuswap and Upper Arrow Lakes, have not yet been examined, but from their similar relation to the Gold Ranges they will probably be found to be equally free from the silt deposit.

In previous publications I have stated the maximum observed height of the main deposits of these silts of the southern part of the Interior Plateau at about 1,700 feet,[2] but later observations show that they are developed to a notable extent at considerably greater elevations. I will here only instance the valleys of Barrière River, a tributary of the North Thompson, Upper Nicola River below Douglas Lake, and Skuh-unh Creek, a tributary of the Lower Nicola, where thick deposits of white silt cut into terraces were observed at heights of about 2,250, 2,500 and 2,450 feet respectively. It is thus probable that we may safely place the upper level of the silt formation in this region at about 2,500 feet, though it is still apparent that the more important developments of the deposit lie below 1,700 feet. This circumstance may be accepted as indicating that the region in question was subjected to elevation of a certain amount during the progress of deposition of the silts, the longest period of rest occurring at about 1,700 feet. The silt deposits are found in this part of the Interior Plateau down to heights less than 1,000 feet, but it is possible that some of the lower-level deposits have been secondarily formed from the denudation of the higher.

In the Nechacco region, situated in the northern part of the Interior Plateau between latitudes 53° 30′ and 54° 30′, the white silt formation is very extensively displayed, covering an area of at least 1,000 square miles, with a thickness of 100 feet in some sections and probably exceeding 200 feet in certain places. The silts reach an elevation of about 2,400 feet at the edges of the basin occupied by them, and where seen lowest (near Fort George) have an elevation of 1,900 feet. It is further possible that certain sandy deposits found near the sources of the Chilacco at a height of about 2,600 feet may

---

[1] The southern portion of this lake has not yet been examined.
[2] 'Report of Progress Geol. Surv. Can.,' 1877-78, p. 143 B; 'Quart. Journ. Geol. Soc.,' vol. xxxvii, p. 275.

represent a littoral condition of the silts.[1] Some remarkably shingly beaches found on the slope to the north of Tsa-whuz Mountain, at a height of about 2,100 feet, were taken to represent a shore-line of the Nechacco white silt lake or sea when at a lower stage.[2] On the north side of the Nechacco basin, the silts go no further than the lower end of Stuart Lake, ceasing there at an elevation of about 2,000 feet, and they are not seen along the upper part of this lake or on Babine Lake. The shores of François Lake, to the west, (2,375 feet) are also free from them.

If not interrupted in some unknown manner, the valley of the Fraser River must at this time have constituted a free connection between the basin of the Nechacco silts and the previously described development of the same silts in the southern part of the Interior Plateau; while to the north-eastward, water standing at the level implied by the Nechacco silts must have extended by way of the Peace River gap through the Rocky Mountain Range.

Though carrying us for a moment beyond the Cordilleran region proper with which we are at present occupied, we may here note that a great area of plateau country to the west of the 117th meridian, which is now cut through by the deep valleys of the Peace and its tributary the Smoky River, is covered with silty deposits precisely similar to those of the Nechacco basin. These generally rest upon the boulder-clay of the region and are embraced between elevations of 2,500 and 2,000 feet.[3]

Though much less important in respect to area, the silty deposits of the Stikine and Tanzilla valleys, between latitudes 58° and 58° 30′, may next be mentioned. These are sufficient to form a wide area of flat country, at a level of about 2,200 to 2,300 feet, and are several hundred feet in thickness.[4] The material is here somewhat darker in colour and more clayey than elsewhere noted, but corresponds in its regular bedding and in the abundance of calcareous nodules which it holds with many typical examples of the white silt formation.

In descending the Dease River, typical white silts are first met with about latitude 59° 30′ at an approximate height above sea-level of 2,400 feet.[5] In the lower part of the Dease Valley, and in that part of the valley of the Liard immediately above the mouth of the Dease, the silts pass into, and appear to be represented by, sandy and gravelly deposits, but on the last-named river near the mouth of the Frances they become again well developed at a height of about 2,300 feet.[6] No opportunity occurred, however, of ascertaining their highest level in this valley.[7]

---

[1] 'Quart. Journ. Geol. Sc.,' vol. xxxiv, pp. 105-107.
[2] 'Quart. Journ. Geol. Soc.,' vol. xxxiv, p. 110.
[3] 'Report of Progress Geol. Surv. Can.,' 1879-80, p. 142 B; 'Quart. Journ. Geol. Soc.,' vol. xxxvii, p. 277. It may be added here, as at least suggestive, that over a great part of the area drained by the various tributaries of the Saskatchewan, at similar elevations, the upper member of the glacial series is constituted by similar silty beds. See 'Report of Progress Geol. Surv. Can.,' 1882-84, part C; and 'Annual Report Geol. Surv. Can.,' 1885, part E; 'Bull. Geol. Soc. Am.,' vol. i, p. 403.
[4] See illustration 'Annual Report Geol. Surv. Can.,' 1887-88. p. 67 B. The level here given is that of the main surface of the deposit, the height of its borders was not precisely ascertained.
[5] 'Annual Report Geol. Surv. Can.,' 1887-88, p. 90 B.
[6] 'Annual Report Geol. Surv. Can.,' 1887-88, pp. 101 B, 103 B.
[7] Though not closely connected with the present discussion of the white silt formation, it may here be noted that the principal terrace found along the shores of Dease and Frances Lakes, (forming the sources of the Dease and Frances branches of the Liard,) were found to be at the level of 3,150 and 3,200 feet respectively. Those lakes are

Crossing finally to the basin of the Upper Yukon, we again find the silt formation well displayed along the valleys of the Upper Pelly and Lewes branches of this great river. Silty deposits were first seen in descending the Upper Pelly at an elevation of about 2,900 feet, but these are of small thickness, and are subsequently interrupted for a considerable distance. Near the mouth of Ross River, however, at an approximate elevation of 2,700 feet, the silts appear in full development with a thickness of fifty feet, and resting indifferently upon the boulder-clay or its local representative the clayey-gravel They continue along the valley from this point for a distance of over 130 miles, and were last seen at a height of about 1,700 feet.[1] From the flat character of parts of the country, they probably extend for considerable distances back from the river in some places.

In ascending the Lewes River, from its point of junction with the Pelly, the white silt formation was first observed about nine miles below the Big Salmon River, at an approximate elevation of 2,000 feet, and extends thence to Lake Marsh or to a height of 2,150 feet. It probably stretches also far up the Teslintoo Valley, which has not been examined. The low elevation of the upward limit of the silts on the Lewes, appears to be explicable on the supposition that the area of their deposition was here limited to the south by the position held by the front of the glacier which occupied the upper part of the valley at the second period of advance of ice. In this case, Lake Marsh and the other long lakes above it, would occupy the same position relatively to the glaciers of the second period as has already been assigned to various lakes in the more southern parts of British Columbia. It seems probable in fact, that these lakes on the north, with Babine, Stuart and Tacla lakes on the south, may be considered as being in direct relation to lobes of ice stretching out on both sides from the region which at an earlier stage had been the central névé of the Cordilleran glacier. The absence of boulder-clay in the upper part of the Lewes valley, which has been particularly referred to by Mr. Russell,[2] I conceive to be due to the same cause i. e. to the circumstance that the glacier during its second advance, swept away whatever deposits had been formed along its foot during the time of its first retreat. Mr. Russell in his paper which has just been alluded to,[3] proposes to name the body of water in which the silt formation was here produced, "Lake Yukon." To this I would only suggest in amendment, in view of the facts set forth in the foregoing pages, together with the circumstance that there is no known natural boundary by which the waters of the supposed lake might have been held back to the north, that it may more appropriately be entitled Yukon Inlet, the presumption being that its waters were in direct communication with the sea.

Before leaving the subject of the white silt formation, which it has been considered advisable to treat consecutively and in some detail, it may be well to tabulate, as follows, the observed levels at which this formation occurs in the various and widely extended localities referred to.

---

distant from each other about 200 miles by the valleys of the rivers by which they discharge into the Liard, and the circumstances are such as to warrant the belief that they were at a time of greater submergence antecedent to that of the deposit of the silts, parts of a single great body of water. This degree of submergence would correspond closely with that required for the formation of the boulder-clay deposits of the Upper Yukon basin, on the northern slope of the Pacific-Arctic watershed.

[1] 'Annual Report Geol. Surv. Can.,' 1887-88, pp. 122 B, 126 B.
[2] 'Bull., Geol. Soc. Am.,' vol. I, p. 143.   [3] Ibid, page 146.

The facts thus summarized apply to a part of the Cordilleran belt 1,200 miles in length, and are I believe sufficient to indicate that the white silts constitute a definite formation, and mark an important period in the history of the glaciation.

TABLE SHOWING THE NORMAL UPPER LEVEL OF THE WHITE SILT FORMATION IN VARIOUS PARTS OF THE CORDILLERAN REGION, WITH SOME OBSERVATIONS ON THE LOWER LIMIT OF THE SAME FORMATION, AND INSTANCES OF LOCAL AND EXCEPTIONALLY HIGH SILT DEPOSITS.

REGION.	Normal highest level, in feet above sea, of main silt formation.	Lowest observed level, in feet above sea, of main silt formation.	Extreme highest level, in feet above sea, of silts locally developed and in some cases evidently due to glacier lakes of small dimensions.
Columbia-Kootanie Valley, (opens to south)	2,700	2,200—	3,500
Southern Part of Interior Plateau, (opens to south and to Pacific)	1,700 & 2,500	1,000—	3,400
Northern Part of Interior Plateau, (opens to Peace River Plains and to south)	2,400	1,900	2,000
Peace River Plains, (drain to Mackenzie River)	2,500	2,000	
Upper Valley of Stikine, (opens to Pacific)	2,200 to 2,300	1,100	
Upper Liard Basin, (opens to Mackenzie)	2,400	(?)	
Upper Yukon Basin, (drains to Behring Sea)	2,700	1,700	2,000

The very numerous observations respecting the white silt formation, which are thus brought together in the form of general statements, have been made by me in connection with the examination of different parts of this extensive region during the past fourteen years, and their inter-relation has only by degrees become apparent. Most of the heights assigned in the above table and on previous pages depend on barometric results, but as these have been carefully checked and are in many cases means derived from a considerable number of readings, they may safely be accepted as sufficiently correct to be employed in a general view of the subject. The bodies of water in which the silts were laid down must in some cases have been bordered by sandy or gravelly beaches, which doubtless passed in depth more or less gradually into true silty deposits; and thus it is that, while the main developments of the silts are striking and apparent, the precise position of their highest margin is not always definitely recognizable, even where the circumstances of observation are otherwise most favorable.

Allowing, however, for these sources of uncertainty, there appears to be a greater amount of difference between the highest levels of the main silt areas than can thus be accounted for. Though generally speaking contemporaneous, it cannot be supposed that the silts of the different areas were absolutely coeval, nor that the changes in level of so vast a tract of the Cordillera were produced with perfect equality. It seems probable, in fact, that the southern part of the Interior Plateau was subjected to a differential elevation of several hundred feet after the main time of deposit of the silts began, while the Columbia-Kootanie valley and the Upper Yukon basin, appear to have stood at an elevation of about 200 feet less than the mean level at the time.

It is unnecessary in this paper to enter into detail respecting the extent of the various more or less local glaciers of the second period, at the time at which their final retreat commenced and the deposit of the white silt formation ended. The distribution of these glaciers conformed closely to the secondary orographic features of the Cordillera, and was thus too intricate to be shown on a small map, and is besides, as yet only partially known. It would appear that a reduced representative of the Cordilleran glacier, probably more or less broken up and divided, remained about the position of the great old névé, while in the southern part of British Columbia the most important local glaciers were connected with the Coast and Gold Ranges, which are to-day the lines of most copious precipitation.

If, in conformity with the opinion of many geologists, the earth's crust may be assumed to respond readily by depression to any considerable local increase of load, we appear to find some ground for a belief that the mere existence of heavy local glaciers about the axial regions of the Gold coast and other ranges, acting throughout the long period which the deposit of the white silt formation implies, may be called in to account for the great depth of the lakes and fiords which occupy the more important valleys in these ranges. It may in fact, on this hypothesis be assumed, that in consequence of a depression affecting the axial parts of such ranges, the pre-existing valleys within their borders, became depressed in their upper parts. Without some such explanation, and unless we are prepared to admit that a greater part of the excavation of these valleys was produced by the wearing action of the glaciers themselves, the existence of some of them is very puzzling. Thus, to cite only two instances,—Adams Lake, in the southern part of the Gold Ranges, is known to be over 900 feet deep in its upper portion, and the Upper Arrow Lake has a depth toward its head of more than 700 feet; depths which in both cases much exceed the probable depth of any buried channel of outflow. The phenomenon here alluded to is, however, one of such frequent occurrence in this part of the Cordillera as to imply some general cause, and is doubtless the same with that which may be found applicable to similar lakes in other parts of the world.[1] The above solution is therefore merely noted as a possible one, and it will be observed, that in exact proportion to the amount to which such axial depression of ranges may be admitted it must be considered as weakening that part of the evidence already drawn from the depths of the fiords of the coast region, in respect to the amount of the later Pliocene and early glacial elevation of the Cordillera.

The final decrease of the ice must apparently have been due to some general cause connected with the close of the Glacial epoch as a whole, for simultaneously with the ultimate retreat or disappearance of the glaciers, the elevation of the Cordilleran region was resumed, til' it eventually reached approximately the level which it now possesses. One feature to which I have called attention on several former occasions appears, however, to require explanation in connection with this final movement in elevation, namely, the almost complete absence of terrace-deposits on the seaward slopes of the Coast Ranges as well as in the valleys of most of the streams and rivers with short courses which drain these seaward slopes or pass through them. This circumstance is believed to depend on

---

[1] It is not the intention here to include all rock basins now containing lakes, but merely those which seem to imply a reversal of slope in mountain-valleys. The fact that differential changes in elevation cannot explain all rock-basins, does not preclude the application of such a hypothesis to special cases. Cf. Prof. J. Geikie in "The Great Ice Age," 2nd edition, p. 275.

the fact that the glaciers of this range were still able to follow the retreating border of the sea, filling the valleys and covering the slopes as they emerged and sweeping before them such detrital deposits as had been formed, their action being necessarily most effective in the valleys and hollows in which, under other circumstances, terraces might most readily have been produced or preserved. We are not, however, entirely without proof, such as may be afforded by detrital deposits, of the elevation of the land at this time; for, in the Puget Sound region, which is more favorably situated for the preservation of these deposits, Prof. Newberry quotes Mr. Bailey Willis as authority for the statement that marine terraces can there be traced to a height of 1,000 feet above the present ocean level.[1]

There is, further, ground for the belief that the movement in elevation of the land was for a time arrested at a stage during which the sea stood about 200 feet higher than it now does. More or less extensive terraces at about this level are found in various places along the coast of British Columbia,[2] and it is to the same period that the more wide-spread deposits of the north-eastern part of the Queen Charlotte Islands are attributed. Local glaciers of considerable size must still have capped the mountains of the Queen Charlotte Islands at this time, leaving on their ultimate disappearance a remarkable series of lake-basins between the base of these mountains and the adjacent submarine deposits of the same period.[3]

In two papers published in the Canadian Naturalist in 1877 and 1878 some observations have been adduced in favour of the belief that the coast region was at one time, immediately succeeding the Glacial epoch and subsequent to the changes above described, re-elevated to an amount of about 600 feet.[4] Facts tending to this conclusion were derived from the inlets now constituting Puget Sound, which are in large part cut out in glacial deposits and have pretty certainly been formed by rivers and afterwards flooded by the sea. This post-glacial epoch of elevation can not yet be considered as having been definitely proved, further information being desirable, but it nevertheless appears to be very probable. If this change be admitted, together with the following subsidence to the present level which a belief in it implies, it is the last of an important character of which any record has been found in this entire region. The occurrence of casts of marine molluscs from which the whole of the carbonate of lime has been removed by the action of subaërial waters, in glacial clays below high-water mark in Embly Lagoon,[5] (lat. 50° 57′), constitutes a slight further indication, which is not unimportant, of a post-glacial elevation of the coast.

One further circumstance may, in conclusion, be referred to here as being readily and intelligibly explicable on the hypothesis of a considerable elevation of the land at about this time. This is the existence at the present day of caribou in the northern part of Queen Charlotte Islands. In a former report on these islands I have spoken of the occur-

---

[1] 'Annals N. Y. Acad. Science,' vol. iii, p. 260.
[2] 'Report of Progress Geol. Surv. Can.,' 1879-80, p. 136 B; ' Annual Report Geol. Surv. Can.,' 1886, p. 103 B.
[3] 'Report of Progress Geol. Surv. Can.,' 1887-88, p. 99 B; 'Quart. Journ. Geol. Soc.,' vol. xxxvii, p. 218.
[4] 'Canadian Naturalist,' vol. viii, (1878), pp. 241, 369. Prof. J. S. Newberry, at a later date, appears to have independently reached an identical conclusion from a study of the same facts. 'Annals N. Y. Acad. Science,' vol. iii, (1884), p. 265.
[5] ' Annual Report Geol. Surv. Can.,' 1886, p. 100 B.

rence of the elk or wapiti on them.[1] This statement was, however, based merely on Indian report, as none of the animals in question were seen. Since that time I have learned from Mr. W. Charles, that the animal in question is really the caribou, and I have been shown by him the skin and antlers of one of these animals. The caribou is not now found anywhere else in the region of the coast, either on the islands or in the Coast Ranges, though it roams over high plateaux to the east of these ranges. The shortest distance between any point of the Queen Charlotte islands and the nearest islands of the Coast Archipelago is thirty miles, and the intervening strait is subject to rapid tidal currents. The isolation of the Queen Charlotte Islands is in fact so complete that the deer, which inhabits all the other islands of the coast, is not found in this group. It is, therefore, in the absence of the caribou from the neighboring coast and its adjacent islands and in consideration of the width of the waterway which would have to be crossed, at least highly improbable that this animal reached the Queen Charlotte Islands under the present conditions. I am thus led to believe that the caribou colonized the islands at a time at which either the glaciers extending from the mainland attained to the Queen Charlotte Islands, or by a land connection during a period of greater elevation.[2] The latter is in every way the more probable supposition, and if it be entertained it may further be assumed that the animal came to the islands at the date of the immediately post-glacial elevation above indicated, and that it has since, as an isolated colony, succeeded in maintaining itself there.

As the height of the northern part of the Cordillera increased, after the period of rest marked by the white silt formation, the streams of the mainland were enabled to begin the cutting out of new channels in the glacial debris with which their valleys had become more or less choked. In consequence of the strongly marked relief of the country, the principal rivers and larger streams were seldom forced by the glacial accumulations to leave their old valleys. Though the amount of this post-glacial river erosion is considerable, it may be regarded as insignificant in comparison with that of the great Pliocene period of denudation. To include any treatment of this subject here would, however, involve appeal to an inadmissible amount of local detail. I will therefore state only, in general terms, that a large proportion of the rivers have up to the present time been unable to cut down to their pre-glacial beds; but that in other places this has been accomplished, and they are now engaged in more or less active erosion of the subjacent rock. On that part of the Fraser River between Lytton and Big-Bar Creek—a length of eighty miles—the post-glacial excavation by the river varies in depth from 400 to 800 feet, in accordance with the evidence of terraces still remaining in the valley. In several places the old channel has been locally departed from, and narrow cañons with depths of 200 feet or more, have been cut across points of solid rock, and in some places (as between Bridge River and Fountain Creek), the Fraser has evidently regained its preglacial level and is now at work in still further deepening the old channel. The conditions met with on the corresponding part of the length of the Main Thompson River are very similar.

Having thus endeavoured to trace in their sequence the events of the Glacial epoch in the Cordilleran region, we may now consider briefly, and with that which has already

---

[1] ' Report of Progress Geol. Surv. Can.,' 1878-79, p. 113 B.
[2] The minimum amount of elevation required would be about 200 feet above the present level.

been ascertained as a basis, the coeval conditions in the adjoining area of the Great Plains. The watershed of the Rocky Mountain Range proper, has always constituted a physical and to some extent also a mental barrier and line of division between our investigations pursued to the east and west of it, and as already pointed out, this is in a measure justified by the small amount of apparent connection or overlap of the evidences of the Glacial epoch of the two regions thus marked out. It is necessary, however, to suppose that the occurrences of this period were to some extent homologous in both areas, or in other words, that if not similar and contemporaneous they were at least correlative.[1]

In the earlier stages of the investigation of the phenomena of the glacial epoch, it was as a rule mentally postulated that in endeavouring to account for these phenomena we were entitled to consider only such changes in elevation as are practically of continental extent, or such as may have affected the northern hemisphere as a whole. Of late years, evidence has been accumulating on all hands, of differential changes in elevation which, in times very recent geologically, have affected comparatively limited areas. I do not propose to enter into any discussion of this evidence, or of the theories by which such changes may be explained, but have found that in admitting their existence, it is possible to arrive at tenable hypotheses leading to reasonably satisfactory conclusions respecting the glaciation of the western part of the continent.

In the history of the northern part of the Cordillera in Mesozoic and Tertiary times, given in the first part of this essay, we have already found indications of the complementary character, from a physical point of view, of the areas of the Cordillera and the Great Plains. There is an antecedent probability in the belief that the elevation of a wide belt of country such as that comprised in the Cordillera, would lead to a correlative depression in some adjacent region, and though this compensating change may here be largely accounted for by supposing the occurrence of upward or downward movements in some not far remote part of the bed of the Pacific, respecting which we are without evidence, it appears to have been affected to a variable and occasionally to a very great degree in the area of the plains.

I would refer particularly in this connection to a suggestive discussion and summary of correlative movements in elevation and depression by Mr. Warren Upham, which appears in the form of an appendix to Prof. Wright's " Ice Age in North America."[2] This discussion appears to hold out the prospect of a solution for many of the difficulties which have so far attended the explanation of the facts of the Glacial period, by putting in a concrete form a series of propositions, which though doubtless present in a more or less recognized manner to the minds of many geologists, have been by them but little employed as implements of research.

Having thus alluded in general terms to the fundamental conceptions involved, I may present the following tentative scheme of correlation of the principal events of the Glacial period in the area of the Cordillera and in that of the Great Plains, which I have

---

[1] Mr. I. C. Russell in his "Notes on the Surface Geology of Alaska," refers to the freshness of glacial striation on Lake Labarge, with other facts, as possible evidence of much greater recency in the glaciation of the Cordillera as compared with that of the eastern part of the continent. 'Bull., Geol. Soc. Am.,' vol. i, p 142. My observations in the northern parts of both regions do not lead me to concur in this suggestion, which in view of all other circumstances seems so improbable.

[2] Appendix A. " Probable Causes of Glaciation," by Warren Upham.

been led to adopt as at least a tenable working hypothesis, after mature consideration of the available evidence relating to both these great areas, and in the light of personal familiarity with the circumstances in both. In giving to this scheme, the definiteness implied by a tabular arrangement, I would, however, premise that though for the purpose of making it clearly understood it is necessary to adopt this form, there is no intention of presenting it as a complete or final solution of the problem under discussion. In view of the multiplicity and more or less indeterminate character of many of the observations which it has been endeavoured to embrace under this generalization, as well as the vast extent and diversity in features of the region to which it relates, it is scarcely possible to hope that it may not eventually prove to be in error in some points. It appears however at present to agree with the known conditions and to indicate at least the *mode* of their explanation.

SCHEME OF CORRELATION OF THE PHENOMENA OF THE GLACIAL PERIOD IN THE CORDILLERAN REGION AND THE REGION OF THE GREAT PLAINS.

Cordilleran Region.	*Region of the Great Plains.*
Cordilleran zone at a high elevation. Period of most severe glaciation and maximum development of the great Cordilleran glacier.	Correlative subsidence and submergence of the Great Plains, with possible contemporaneous increased elevation of the Laurentian axis and maximum development of ice upon it. Deposition of the lower boulder-clay of the plains.
Gradual subsidence of the Cordilleran region and decay of the great glacier, with deposition of the boulder-clay of the Interior Plateau and the Yukon Basin, of the lower boulder-clay of the littoral and also at a later stage (and with greater submergence) of the interglacial silty beds of the same region.	Correlative elevation of the western part of the Great Plains, which was probably more or less irregular and led to the production of extensive lakes in which inter-glacial deposits, including peat, were formed.
Re-elevation of the Cordilleran region to a level probably as high as or somewhat higher than the present. Maximum of second period of glaciation.	Correlative subsidence of the plains, which (at least in the western part of the region) exceeded the first subsidence and extended submergence to the base of the Rocky Mountains near the forty-ninth parallel. Formation of second boulder-clay, and (at a later stage) dispersion of large erratics.
Partial subsidence of the Cordilleran Region to a level about 2,500 feet lower than the present. Long stage of stability, during which the *White Silts* were laid down. Glaciers of the second period considerably reduced. Upper boulder-clay of the coast probably formed at this time, though perhaps in part during the second maximum of glaciation.	Correlative elevation of the plains, or at least of their western portion, resulting in a condition of equilibrium as between the Plains and the Cordillera, their relative levels becoming nearly as at present. Probable formation of the Missouri Côteau along a shore-line during this period of rest.
Renewed elevation of the Cordilleran region with one well marked pause, during which the littoral stood about 200 feet lower than at present. Glaciers much reduced and diminishing, in consequence of general amelioration of climate toward the close of the Glacial period.	Simultaneous elevation of the Great Plains to about their present level, with final exclusion of waters in connection with the sea. Lake Agassiz formed and eventually drained toward the close of this period. This simultaneous movement in elevation of both great areas may probably be connected with the more general northern elevation of land at the close of the Glacial period.

Referring to the several correlative movements of elevation and of depression of the Cordillera and the Great Plains implied by the hypothesis set forth in the above comparative scheme, it may I think be admitted as not improbable on theoretical grounds, that such conditions of oscillation once set up, in consequence of the interaction of whatever causes, might have a tendency to continue for a considerable time before a stable condition was regained ; a state of equilibrium being in the end attained either by the general decrease in intensity of the forces involved or by the final preponderance of one class of these. It may further be pointed out, that the supposed sequence of events is generally in correspondence with the view that the periods of maximum glaciation of the Cordillera were those of its greatest elevation, while the decay of its great glaciers was in both instances accompanied, if not caused, by subsidences leading to the encroachment of oceanic waters. The flooding of the Great Plains by arctic water, while the Cordillera stood as a much elevated land between these and the warmer waters of the Pacific, would in itself go far to explain the conditions under which the excessive precipitation required for the production of the Cordilleran glacier occurred.

The supposed sequence of events is furthermore compatible with the belief, which has been advanced by several geologists, that the weight of a ponderous ice-cap is in itself sufficient to produce subsidence of the land, and with the idea that such an ice-cap may thus eventually become self-destructive, by obliterating the elevation to which its existence is in the first instance, largely due.

It is also of interest, though as a minor point, to note that the extensive occurrence, in the interglacial beds in the area of what is now one of the most arid parts of the north-western plains, of peaty deposits abounding in the remains of ferns and semi-woody plants,[1] may be accepted as betokening a considerable local rainfall ; a circumstance which would be in accord with the postulated depression of the Cordillera and relatively elevated position of the Great Plains at that time.

Though independently based upon and primarily intended to include the observed phenomena of the Glacial period in the north-western portion of the continent alone, it is further worthy of remark, that the elevation believed to have affected the Canadian Great Plains during the interglacial episode, is to some extent confirmed by the fact that Messrs. Chamberlin and Salisbury find evidence of a similar elevation of the Upper Mississippi Valley, at a corresponding time, to an extent of about 1,000 feet.[2] It is also noteworthy that the two great correlative movements of elevation of the Cordillera and depression of the Great Plains here admitted, correspond, at least in a general way, to the principal similar, though smaller, changes in level accepted by Mr. W. J. McGee as explaining occurrences on the "Middle Atlantic slope," as indicated by the Columbia formation.[3] As in the case of the Great Plains, the two marked changes in the region of the Middle Atlantic slope were in the sense of depression, but as there is reason to believe that depression in one region may have been made up for by elevation in another, the main point to which allusion is here made, is merely that both maxima of glaciation appear to have been marked on the Pacific as well as on the Atlantic side of the continent, by considerable disturbance in level.

In previous publications on the glaciation at the Canadian portion of the Great Plains,

---

[1] "Bull. Geol. Soc. Am.," vol. I, p. 332.
[2] "Sixth Annual Report U. S. Geol. Surv.," p. 214.
[3] "Seventh Annual Report. U. S. Geol. Surv.," p. 639.

dealing only with the evidence available at the time, which did not include any considerable body of facts relating to the Cordillera, I had arrived at the opinion that, as a whole, this glaciation was attributable to the action of water-borne ice rather than to that of the ice of a great confluent glacier. At the same time in deference to the views maintained by many geologists, I pointed out in what manner the circumstances might be explained under either hypothesis and noted the difficulties incident to both.[1] At this time I fully recognized the fact that the Laurentian axis had been buried beneath a confluent glacier, and the evidence upon which the existence of a similar confluent Cordilleran glacier is based, as noted in previous pages, has since been obtained.[2] Later and more detailed and systematic examinations of the area of the Great Plains, carried out with a knowledge of the views of other geologists on the glaciation of different parts of the northern hemisphere, have, however, resulted only in further confirming my belief in the glacio-natant origin of the glacial deposits of these plains. I may be excused for stating here that I have had every opportunity of becoming familiar with the effects of confluent glacier action in regions to the east and west of the plains, my apology for making this statement being found in the circumstance that the attribution of all the widespread and important phenomena of the Glacial period to the action of a confluent glacier (a lineal descendant of the polar ice-cap) has become almost official; with the natural if scarcely conscious tendency, which for a time appears to follow the acceptance of any wide generalization, to ignore or doubt any outstanding or unconformable observations rather than to appreciate these at their full value and to follow them up as useful clues toward a position of fuller knowledge of the phenomena as a whole. Some reasons for the opinion held by me that the effects met with on the plains generally must be accounted for by floating ice, are alluded to on a later page.

It may here be noted, however, that the theoretical mode of accounting for the glaciation of the Great Plains which appeared to me most probable at an earlier date, and which was discussed in the publications just alluded to, has since become subject to considerable modifications, principally by reason of the discovery of boulder-clays of two distinct periods and the definition of certain intervening interglacial deposits. Here, as in the Cordilleran region, the existence of evidence of the action of water at high levels might be simply referred to the occurrence of a single and contemporaneous time of great depression of the land, but though the existence of such evidence at heights of between 4,000 and 5,300 feet in the Cordillera and on the eastern slopes of the Rocky Mountains, may be referred to as favoring the theory of a general depression of this kind, it is

---

[1] "Geology and Resources of the Region in the vicinity of the Forty-ninth Parallel" (1875). 'Quart. Journ. Geol. Soc.,' vol. xxxi (1875).

[2] It being now definitely known, as a result of work done under the auspices of the Geological Survey of Canada and more particularly of observations by myself and my colleagues Messrs. McConnell and Tyrrell, that the extreme borders of the eastern and western glaciated areas of the continent barely overlap, and then only in a limited region; it becomes manifestly necessary to recognize two distinct main centres of glaciation, both of which were, at the maximum of the glacial conditions, more or less completely covered by great confluent glaciers. Cf. 'Geological Magazine,' Dec. iii, vol. v, p. 250. It is therefore no longer appropriate to refer to the great confluent ice-mass which spread out from the region of the Laurentian axis or plateau as the "Continental Glacier," even if it be supposed that it extended as far as the western edge of glacial deposits on the plains. In view of this fact and in the interest of precision of expression, I venture to suggest that this may be named the *Laurentide Glacier*. For the great ice-mass of the western region I have already proposed the name *Cordilleran Glacier*.

believed that the proof of the high level of the Cordillera and the relatively low elevation of the plains at the initial stage of the Glacial period, constitutes an insuperable difficulty in the way of a simple explanation based on such a supposition.

As the scope of this paper does not include any extended description of the glacial phenomena of the Great Plains, I will here only outline, in the most general terms possible, the character of the observed facts [1]

The surface of that part of the Great Plains which is characterized by deposits due to the Glacial period, is at the present time highest at the base of the Rocky Mountain Range near the forty-ninth parallel, having there an elevation of about 4,000 feet above the sea, and sloping thence in a north-eastward direction to the base of the Laurentian axis or plateau on the east (distant about 700 miles), with an average fall of about 5·4 feet to the mile. Further north, in the Peace and Liard River countries, the width of the region representing the plains between the Rocky Mountains and the Laurentian plateau is much less, but it still preserves an easterly slope, the highest plains to the east of the foot-hills of the Rocky Mountains being at an elevation of about 3,000 feet in the first mentioned district and still lower in the second. The maximum elevation attained by the Laurentian plateau in the corresponding part of its length is under 2,000 feet, and probably averages about 1,600 feet only, and from the rocks of this plateau a considerable portion of the whole material of the drift, even of the highest levels, has been derived.

The glacial drift reaches quite to the foot of the Rocky Mountains, near the forty-ninth parallel, but soon leaves them to the south, while further to the north its margin is constantly found at a somewhat lower level and at some distance from the base of the mountains. Disregarding exceptional areas of small dimensions, from some of which the glacial drift has been more or less completely removed by denudation, and, in a few cases, upon which it has never been deposited, the whole of this vast tract is covered with remarkable uniformity and with little regard to the elevation by deposits referable to the boulder-clay. Overlying these, stratified sands or gravels are almost everywhere found in continuous and connected spread, with, very frequently, bedded silty deposits; some of which are evidently due to post-glacial lakes, others apparently to more general causes. Large erratics are in all parts of the region observed to be most abundant on the surface, or in connection with some of the upper superficial deposits.

In the western part of the plains, but so far as ascertained to the east of the extreme western margin of the drift deposits, the boulder-clay is found to be separable into two parts, between which interglacial silty and sandy deposits, which in some places hold peat, are extensively developed. It is as yet uncertain whether the division thus evidenced as between the lower and upper boulder-clays extends to the eastern half of the plains, the area over which it is known to be well marked, stretching from the vicinity of the forty-ninth parallel northward to the North Saskatchewan River, with a length of about 250 miles and a maximum actually known width of about fifty miles.

---

[1] For details of observation in this area reference may be made particularly to "Geology and Resources of the Forty-ninth Parallel," page 263, 'Quart. Journ. Geol. Soc.,' vol. xxxi, p. 603, 'Report of Progress Geol. Surv. Can.,' 1879-80, p. 136 B., 'Report of Progress Geol. Surv. Can.,' 1882-84, p. 139 C., 'Annual Report Geol. Surv. Can.,' 1885, p. 71 C. (R. G. McConnell), 'Annual Report,' 1886, p. 139 E. (J. B. Tyrrell); also to 'Bull. Geol. Soc. Am.,' vol. I., p. 395 (J. B. Tyrrell), and p. 540 (R. G. McConnell.)

The relation of the lower boulder-clays to certain underlying and immediately pre-glacial gravels has been referred to on a former page (see p. 22.) In accordance with the general hypothesis here advanced, it may be supposed that the submergence of the plains during the first period of glaciation (that of the maximum of the Cordilleran glacier) extended as far westward as the limit of the lower boulder-clay. Between the forty-ninth and fiftieth parallels it must in this case have reached to a position forty miles or less from the base of the Rocky Mountains, and further north to an undetermined line, which, however, in all cases appears to have fallen short of the maximum westward spread of the glacial drift.[1]

The interglacial episode seems to have been marked by the elevation of the western part of the Great Plains, accompanied by the production of shallow lakes, in or around the borders of which peaty deposits were locally formed. The width of the region raised above the sea at this time also remains indeterminate, and no evidence of this episode has yet been obtained to the north of the North Saskatchewan. It appears probable, if the region of considerable uplift extended much further to the northward, that it may have been there narrower and less important. In connection with the description of the Missouri Côteau, I have noticed the existence of systems of now more or less completely abandoned old valleys cut in the Laramie plateau of the vicinity of Wood Mountain.[2] These have been explained as pre-glacial valleys, but in view of the additional light on the sub-division of the Glacial period since obtained, it is now believed that the greater part of the erosion of these wide old valleys in the higher portions of the Great Plains, both in the locality here particularly cited and in the Cypress Hills Plateau and elsewhere, may probably be assigned to the interglacial time of elevation. This would accord better also with their fresh and unbroken outlines, which, in attributing them entirely to pre-glacial action, are difficult to understand.

Before leaving the subject of the interglacial deposits of the western part of the plains, it should be pointed out that, while they indicate that this was in part a land-surface at the time of their deposit, they also afford a certain amount of definite evidence showing that at the date of their formation, this part of the plains had not had impressed upon it the slope from west to east which it now possesses. This is shown particularly along the Belly River, where interglacial beds evidently formed in a shallow lake, may be traced continuously for forty-five miles from west to east. These beds may safely be assumed to approximately represent a water-level of the time, but their position is now no longer horizontal, but conforms instead to the slope which the plains now possess.[3]

To the renewed subsidence of the Great Plains which is supposed to have been contemporaneous with the second elevation of the Cordillera, and correlative to this, is attributed a second access of floating ice, derived chiefly from the front of the confluent glacier of the Laurentian axis (great Laurentide glacier), to which the deposition of the upper boulder-clay is due. We are as yet unable to form any definite opinion as to the extent to which the action of such floating ice and the continued deposition of boulder-clay was limited in the eastern part of the plains during the interglacial episode.

It would further appear, that while the elevation of the Cordilleran belt was probably

---

[1] i.e., The available evidence at least does not carry it so far.
[2] "Geology and Resources of the Forty-ninth Parallel," p. 230. Quart. Journ. Geol. Soc.,' vol. xxxi, p. 615.
[3] 'Report of Progress Geol. Surv. Can.,' 1882-84, p. 151 C.

not greater and possibly not as great as during the first period, the extent to which the western part, at least, of the plains was depressed was in excess of that by which they had been affected before. Various conjectural explanations of this might be hazarded, but in the absence of any complete knowledge of the circumstances, these would possess little value. It would appear, however, that the ice-bearing water at this time reached quite to the base of the Rocky Mountains near the forty-ninth degree of latitude, bringing with it debris derived from the Laurentian axis, of which the nearest part is over 500 miles distant.[1] It does not of course necessarily follow that such debris was actually derived from the nearest possible source, though in previous publications I have spoken of it as thus originating in order to simplify the conceptions involved. Recent investigations by Mr. J. B. Tyrrell, have in fact shewn, that the direction of movement of the ice over the eastern part of the plains may generally have been from north to south or from north-east to south-west,[2] which might imply an even greater distance of carriage for the erratics here particularly alluded to.

Many of these erratics along or near to the base of the mountains between the forty-ninth and fiftieth parallels, lie at heights exceeding 4,000 feet, while the highest observed instances of their occurrence, are at an elevation of 5,280 feet above the present sea-level,[3] the erratics being here stranded upon moraine ridges due to local glaciers which have flowed out from the valleys of the Rocky Mountains, probably during the first maximum of glaciation.

From the same south-western portion of that part of the plains which is characterized by glacial deposits, further evidence, shewing that the elevation succeeding the second subsidence was unequal in character, is afforded by the difference in heights of the upward limit of Laurentian material in a few conspicuous instances. Thus, taking a belt of country running eastward from the base of the mountains and approximately included between the forty-ninth and fiftieth parallels, we find the furthest western erratics at the great elevation just quoted. Seventy-five miles to the east, the summit of Rocky-Spring Plateau is driftless, the upward limit of the eastern drift being found at 4,100 feet. Twenty-five miles further east, the limit reaches a height of 4,660 feet or possibly slightly more on the Three Buttes or Sweet-Grass Hills.[4] Fifty miles still farther to the east, Mr. McConnell has ascertained the limit to be at 4,400 feet on the Cypress Hills, the higher part of the plateau so called being again driftless.

One hundred and sixty-five miles due north of the Rocky-Spring Plateau, Mr. Tyrrell has found the summit of the Hand Hills to be similarly without glacial debris, the limiting height at this place being 3,400 feet.[5]

The localities thus enumerated include all the known driftless areas within the western border of the drift-covered area of the plains, and it is probable that there are not any other instances remaining to be discovered.

---

[1] As indicated on the map which accompanies my first paper on the glaciation of the plains, the actually nearest part of the edge of the Laurentian plateau, on a north-east bearing, is about 550 miles distant. The distance of 700 miles given in that paper, refers to that part of the Laurentian margin between E. and E. N. E., from which alone it was at the time (in conformity with indications of direction obtained from striæ in the Laurentian region), supposed that the travelled material of the part of the plains in question was derived.

[2] 'Bull. Geol. Soc. Am.,' vol. i, p. 401.

[3] 'Report of Progress Geol. Surv. Can.,' 1882-84, pp. 146 C, 148 C.

[4] 'Report of Progress Geol. Surv. Can.,' 1882-84, p. 148 C.

[5] 'Annual Report Geol. Surv. Can.,' 1885, p. 75 C.

It is worthy of remark that the Three Buttes or Sweet-Grass Hills form an isolated case of the occurrence of volcanic rocks in this part of the plains, representing the stumps of volcanoes which are referred to the Miocene period (see p. 15), and occupying the centre of a wide low anticlinal, the influence of which has now been traced northward to the North Saskatchewan. The greater amount of elevation by which the upper limit of the glacial deposits has been affected in the central region of this anticlinal swell, as compared with the instances to the east and west of it, together with the still greater height to which the same limit has been lifted on the flanks of the Rocky Mountains in the same latitude, is very suggestive of the influence of tangential pressure in the earth's crust, which has had most effect in increasing the elevation of old lines of uplift. Rocky-Spring Plateau and the Hand Hills may be considered as about on the nodal line between this anticlinal and the wide synclinal which separates it from the base of the Rocky Mountains.

The depth of submergence of the general surface of the plains surrounding the Cypress Hills and Three Buttes may thus be supposed to have averaged about 1,500 feet, and to have exceeded this by 500 feet or more at a distance of about fifty miles to the north where the general height of the country is less.[1]

Following the western margin of the boulder-clay to the northward, we find it to occur at lower elevations and at a more or less considerable distance from the base of the Rocky Mountains. About the fifty-first parallel its height is from 3,400 to 3,300 feet, at a distance of thirty-five miles from the mountains, while still further to the north it crosses the Red Deer and North Saskatchewan Rivers at an elevation of about 3,000 feet, approximately following a contour-line at that height. Scattered Laurentian boulders are, however, found at somewhat higher levels and to the west of the recognized margin of the boulder-clay up to 3,200 and 3,400 feet.[2] Beyond the North Saskatchewan, the country toward the base of the mountains has not been so closely examined as to ensure that the highest western levels of the Laurentian drift have been accurately determined. On the watershed between the Athabasca and Peace Rivers, however, (lat. 54° 12', long. 117°), Laurentian boulders are found at a height of 3,300 feet. In the Peace River valley, such eastern erratics have been noted to heights of 2,300 to 2,500 feet only, (in the vicinity of the D'Echafaud River in latitude 55° 45', longitude 120°,) being there at a distance of about seventy miles from the eastern base of the mountains proper.[3] According to recent observations by Mr. McConnell, Laurentian boulders are again found in the Liard River region to occur some distance to the west of the margin of the boulder-clay, their greatest observed height being 2,300 in lat. 60° 15' long. 123°.[4]

In the tabular comparison of the events of the Glacial period in the regions of the Cordillera and the Great Plains, it is suggested that the remarkable monument of this period known as the Missouri Côteau,[5] may with probability be referred to that time of

---

[1] Mr. Upham accepts the observations here referred to as evidences of the thickness of a confluent glacier which he supposes to have swept across the plains to the base of the Rocky Mountains. 'Am. Geol. Magazine,' vol. iv, p. 215. For reasons elsewhere stated, I am unable to concur with him in this hypothesis.
[2] 'Annual Report Geol. Surv. Can.,' 1886, p. 143 E, 'Bull. Geol. Soc. Am.' vol. l, p 397.
[3] 'Report of Progress Geol. Surv. Can.,' 1879-80, pp. 130 B, 140 B, 'Quart. Journ. Geol, Soc.,' vol. xxxvii, p. 277.
[4] 'Bull. Geol. Soc. Am.,' vol. i, p. 542. The boulder-clay in this region would appear to be confined to levels below a height of about 1,000 feet.
[5] This natural feature was noted and named by the early French voyageurs *Grand Côteau de Missouri*.

rest and stability which is marked in the Cordillera by the white silt formation. This period seems to have been initiated by a partial subsidence of the Cordillera, accompanied by a correlative elevation of the western part at least of the Great Plains, during which or as a consequence of it, a condition of equilibrium was established as between the Cordillera and the plains, which practically closed the series of correlative oscillations which it is supposed were contemporaneous with the conditions of glaciation in the west. It is of course possible that the Côteau may have been produced along the western margin of the interior continental sea during the elevation by which the plains were affected in the interglacial episode, but so far no evidence with this tendency has been obtained, while the slight degree to which the irregular ridges and mounds of the Côteau have since been affected by denudation, constitutes in itself an argument in favour of a later date.

North of the forty-ninth parallel, the thick and irregular drift deposits of the Côteau, forming a marked feature on the plains, and running in a north-west by south-east bearing, have been examined in some detail for a length of about 200 miles, or to the elbow of the South Saskatchewan.[1] Throughout this part of its length the deposits forming the Côteau are piled on or against a more than usually abrupt slope of the surface of the plains. To the north of the South Saskatchewan, the Côteau is probably more diffuse, in conformity with the less marked edge of the third steppe of the prairie, but it appears to reach the North Saskatchewan near its elbow. Its material consists largely of boulder-clay, but this is generally covered with stratified sands and gravels and strewn with many erratics.[2] Its height is nearly uniformly maintained at from 2,000 to 2,200 feet above the present sea-level, and from its rougher and higher parts it subsides toward the north-east into wide swelling undulations, which slope gradually down to merge into the general level of the plains of the second prairie steppe. The whole appearance of this remarkable glacial deposit is to my mind that of a zone against which heavy debris-bearing ice stranded for considerable period during which no great change in level occurred. When, in 1875, I first recognized the glacial origin of the Côteau, it was pointed out that it might be explained either as the moraine of a continental glacier or in the manner just outlined, but subsequent investigation appears to me to add further probability to the hypothesis that it was formed along a sea-margin. The line of the Côteau is practically concentric, for a considerable part of its length, with that of the outer and higher border of the glacial deposits of the Great Plains; and in this connection it is further worthy of remark that practically the entire area of the plains which is characterized by glacial deposits lies on the arctic slope of the continent, thus rendering it probable that the water by which it appears to have been flooded, was in direct communication with the Arctic Ocean and may in fact be regarded as having been an expansion from this northern sea. It may be added, that in assuming the Côteau to represent a water-line, of which the relative elevation of different parts has since suffered little change, we are led to believe that the line of maximum slope by which the surface of this part of the Great Plains was affected (which slope was evidently increased subsequently to the formation of the Côteau) must lie at right-angles to its length, or from north-east to south-west, a belief which accords precisely with the actual great height of the south-western portion of the drift-covered area.

---

[1] "Geology and Resources of the Forty-ninth Parallel," pp. 228, 256; 'Quart. Journ. Geol. Soc.,' vol. xxxi, p. 614; 'Annual Report Geol. Surv. Can.,' 1885, part C.

[2] 'Annual Report Geol. Surv. Can.,' 1885, p. 74 C.

It has already been stated that there is no proof that any part of the great Cordilleran glacier at any time flowed eastward to the plains by the passes of the Rocky Mountains. The entire absence, whether in the Rocky Mountains or on the plains, of material derived from the Gold Ranges, is alone almost sufficient to show that nothing of the kind happened. Local glaciers of considerable size were, however, produced in these mountains, and such glaciers, debouching to the eastward, formed moraines among the foot-hills and even beyond these on the margin of the plains. Moraines due to these glaciers have been traced, in a more or less degraded condition, for about ten miles beyond the base of the mountains near the forty-ninth parallel. A glacier of larger dimensions than most of those derived from the mountains, evidently followed the Bow Valley for a distance of at least twelve or fourteen miles beyond the mountains.¹ These large local glaciers are believed to have been contemporaneous with the maximum development of the Cordilleran glacier. The material of their moraines has been derived entirely from the Rocky Mountains, but Laurentian erratics are occasionally found resting upon them, and they are frequently surrounded by stratified deposits of later date. Some evidence has already been noted which tends to prove that the Rocky Mountain glaciers were insignificant during the second maximum of glaciation of the Cordillera. (See p. 45.)

The correlative and complementary elevation and depression of the Cordilleran belt and the Great Plains which has been investigated in foregoing pages, necessarily implies either the existence of extensive contemporaneous faulting along the north-eastern margin of the Cordillera, or a hinge-like movement about the same line, the difference in elevation, in the latter case, being accounted for by slopes of varying degrees of inclination. In the absence of any evidence of such recent faulting on a large scale, the latter remains the more probable hypothesis. If the sequence of events which has been advanced be in the main correct, the greatest change in height between neighboring tracts must have occurred during the second period of glaciation, in the southern part of the region under discussion, and may have amounted to between 5,000 and 6,000 feet. There appears to be, however, even here, ample room in which to distribute the resulting slope. Under the assumption that the difference occurred between the line of the Columbia-Kootanie valley and the extreme western margin of the eastern drift, a slope of one in sixty would require to be impressed on the intervening tract, while, even if it be assumed that the whole of this difference occurred between the axis of the Rocky Mountains and the same eastern-drift margin, this slope would only be about doubled, and would thus amount to about 200 feet to the mile.²

The last stage of the Glacial period appears to have been marked by a general movement in elevation of both the Great Plains and the Cordillera, which it is presumed may have been connected and contemporaneous with a still more general movement in the same sense, by which the entire northern part of the continent was affected. Mr. Upham's researches in connection with the later-glacial or post-glacial "Lake Agassiz," which occupied the valley of the Red River and the basins of Lake Winnipeg and associated lakes, proves that in the eastern part, at least, of the plains, this elevatory move-

---

¹ 'Report of Progress Geol. Surv. Can.,' 1882-84, p. 145 C.

² Considerable relative changes in elevation between the western and eastern sides of the Rocky Mountain Range may be found to explain some puzzling features of the actual drainage system of this range. See 'Annual Report Geol. Surv. Can.,' 1885, p. 27 B.

ment was greatest to the north. The north-eastward slope of the westward part of the plains may very probably have been further increased at this time.

In the "Geology and Resources of the Forty-ninth Parallel" facts are adduced tending to shew that a relative elevation of the southern part of the western portion of the plains is the latest change of this kind of which any record is found.[1] The evidence of this is derived chiefly from alterations in the courses of rivers to more northerly directions, and such evidence has since been much added to and extended so as to include nearly the entire area of the Districts of Alberta and Assiniboia.[2]

This elevation of the southern, or relative depression of the northern part of the plains may with great probability be assumed to have immediately succeeded the change in an opposite sense last mentioned. To the west of the Rocky Mountain Range, similar evidence of southward elevation or northward depression in immediately post-glacial time, is afforded by the Columbia-Kootanie Valley,[3] and it is even possible that the outflow of the Great Shuswap Lake became changed at the same time, from the Spallumsheen and Okanagan Valley to that of the South Thompson.[4] The observations here alluded to are at least sufficient to show that a depression to the north or elevation to the south was very widespread in the western part of the continent.

In view of the fact that an undoubted majority of the geologists who have interested themselves in the phenomena of the period of glaciation in America are at present disposed to attribute these phenomena almost in their entirety to the action of one or more great confluent glaciers, while my study of the glaciation of the northern part of the Great Plains leads me to hold the belief that this region has as a whole been submerged, and that floating ice has been the main agent in its glaciation, it appears to be necessary to add here, by way of justification for this belief, a synopsis of the principal facts upon which it is based. This must, however, necessarily be brief and in some measure imperfect, as in the case of many circumstances observed on the ground, sufficiently precise data have not yet been obtained to enable definite statements to be made respecting them in a summary form.

Prof. J. S. Newberry has stated that "the track of a glacier is as unmistakable as that of a man or a bear, and is as significant and trustworthy as any other legible inscription."[5] I would propose to extend this forcible statement of the case by adding that the evidences of the action of glacier-ice and floating-ice are as different as are the tracks of a man from those of a bear, though cases may nevertheless occur in which it is impossible to decide to which agent certain traces should be assigned. Proof of the existence of a great confluent glacier on the Laurentian axis, and of a second similar glacier on the Cordillera, appears to me to be complete, while over the greater portion of the extent of the northern part of the plains the evidence seems to have a different meaning.

One of the most striking circumstances in favour of a belief in the sub-aqueous origin of the glacial deposits of the plains, is their extraordinary persistence and similarity in character over the vast area which they cover in an almost uniform sheet. The area of

---

[1] *Op. Cit.* p. 264.
[2] 'Report of Progress Geol. Surv. Can.,' 1882-84, pp. 14 C, 150 C; 'Annual Report Geol. Surv. Can.,' 1885, p. 75 C; 'Annual Report Geol. Surv. Can.,' 1886, p. 146 E.
[3] 'Annual Report Geol. Surv. Can.,' 1885, p. 31 B.
[4] 'Report of Progress Geol. Surv. Can.,' 1877-78, p. 60 B. [5] 'Geol. Surv., Ohio,' "Geology," vol. ii, p. 2.

that portion alone of this region which is included between the forty-ninth and fifty-fourth parallels of north latitude, exceeds 250,000 square-miles, being about four times as great as the total area of the New England States. Over this entire tract, the boulder-clay and other overlying deposits form a nearly unbroken superficial layer of almost infinitesimal thickness in comparison to its extent, the occurrence of ridges or local accumulations like those of the Côteau and others not here particularly cited, though notable and important from certain points of view, being as compared to the whole exceptional and rare.

Though true that the glacial deposits, and more particularly those of the boulder-clay, are usually found in greatest thickness in less elevated parts of the plains, and exhibit besides a distinct general tendency to diminish in thickness from east to west,[1] this can scarcely be considered as affecting the main statement just made, and may be explained under the hypothesis of deposit either by a great glacier or by water-borne ice. The evidence which has been adduced to shew that the western part of the plains was for a shorter time and less continuously submerged than the eastern, appears to me, however, to account for this general fact most naturally. The transport of a great part of the material of the glacial deposits from east to west or from north-east to south-west is shown by its character, which is such that it can not have been derived elsewhere than from some part of the Laurentian axis. In assigning the origin of the Laurentian material of the western verge of the drift to the nearest possible source (irrespective of views respecting the precise direction of its journey) we find that it must have been transported for about 550 miles.

The boulder-clay of the plains, besides its far-travelled constituents, always includes a considerable and often a large proportion of relatively or quite local material. Owing to the imperfectly consolidated character of the Cretaceous and Laramie deposits of the plains, this important local element is found most abundantly as a portion of the paste, which varies in color to a considerable extent in correspondence with that of neighboring strata.[2] Notwithstanding the observable diversity thus caused, the composition of the boulder-clay throughout, particularly in reference to the constant proportion of far-travelled material, may be characterized as extremely uniform.

More or less apparent signs of stratification are generally to be met with in sections of the boulder-clay of the plains, and in some cases interbedding of different varieties of material, including layers of gravel, sand, etc., are quite prominent features and of such a character as clearly to require the action of water. Where the lower and upper boulder-clays are distinguishable, the evidences of bedding are usually most conspicuous in the latter.[1] The interglacial deposits of the western part of the plains form, as already indicated, an extensive and regular layer between the lower and upper boulder-clays, showing apparently continued deposition at the time which they represent, though under different

---

[1] The greatest known thickness of drift deposits (which are not, however, all boulder-clay) occurs in a well at McLean Station on the C. P. Ry., and is 495 feet, or more. 'Trans. Royal Society of Canada,' vol. iv, Section iv, p. 92. The greatest observed thickness of the boulder-clay, embracing both upper and lower divisions, may be stated at about 200 feet.

[2] 'Report of Progress Geol. Surv. Can.,' 1882-84, p. 143 C; 'Annual Report Geol. Surv. Can.,' 1885, p. 71 C; 'Annual Report Geol. Surv. Can.,' 1886, p. 142 E.

[3] "Geology and Resources of the Forty-ninth Parallel," pp. 239, 240, 242; 'Report of Progress Geol. Surv. Can.,' 1882-84, p. 143 C; 'Annual Report Geol. Surv. Can.,' 1885, p. 72 C; 'Annual Report Geol. Surv. Can.,' 1886, p. 141 E.

circumstances. The stratified deposits of sand, gravel and silty material which still nearly everywhere overlie the boulder-clay, though in some cases no doubt in part referable to post-glacial lakes of limited dimensions, are too wide-spread and are found at too many different levels to allow this to be accepted as a general explanation of their occurrence. Irregularly stratified deposits of this kind even cover most of the hills and ridges of the Missouri Côteau, respecting a part of which Mr. McConnell states, that while the present surface undulations appear to be largely due to irregularities in the old boulder-clay floor, the modified drift wraps both hill and hollow in a nearly uniform sheet, although usually thickening somewhat in the depressions.¹ In the southern part of the District of Alberta, these superficial deposits are very generally found to consist of gravels and sands below, with sandy or clayey loam above.

That the overlying well-stratified deposits have not been formed by the denudation and rearrangement of the boulder-clay alone, is strikingly evidenced by the fact that the largest far-travelled erratics are usually found to be associated with these deposits, lying upon the boulder-clay or even resting on the higher well-stratified materials and projecting conspicuously on the surface of the plains.² This has now been observed to be the characteristic mode of occurrence of the larger boulders over almost the entire area of the Great Plains, where it may frequently be noted that no boulders imbedded in the boulder-clay of the vicinity equal the superficially scattered erratics in size. Mr. McConnell may again be quoted in this connection as follows:—" In many parts of the district, and more especially on rough ridges like the Côteau, the surface is almost covered with gneissic and limestone boulders and angular fragments, which appear at first sight to have been sifted out of the drift below by denudation, and this is, no doubt, true in regard to a large proportion of them. Many of these erratics are, however, much larger, and are also less water-worn than they would be if derived from that source, and it is highly probable that these represent a more recent period of distribution."³ In further amplification of this point the various reports which have already been referred to may be consulted.⁴

It is probable that the boulders found scattered over the more prominent isolated plateaux on which little other trace of glacial deposits is found, as well as the highest erratics met with on the Sweet-Grass Hills and those stranded at great elevations on moraines along the base of the Rocky Mountains, are referable to the same final period of distribution. The remarkable group of large erratics composed of Huronian quartzite (and evidently derived from some occurrence of this rock in the eastern Laurentian highlands,) which is found near the lower part of the Waterton River, at a distance of only thirty miles from the base of the Rocky Mountains and at a height of 3,200 to 3,300 feet, belongs to this later period of dispersion. One of these erratics is forty-two and another forty feet in length.⁵ The occurrence of a group of such large boulders of *identical and exceptional* lithological character, resting upon the surface of the prairie in a single locality

---

¹ ' Annual Report Geol. Surv. Can.,' 1885, p. 74 C.

² It is further remarkable that some of the largest eastern erratics are found near the western limit of the drift. ' Report of Progress Geol. Surv. Can.,' 1882-84, p. 148 C.

³ ' Annual Report Geol. Surv. Can.,' 1885, p. 74 C.

⁴ See particularly "Geology and Resources of Forty-ninth Parallel," pp. 218. 255; ' Quart. Journ. Geol. Soc.,' vol. xxxi, p. 611; ' Report of Progress Geol. Surv. Can.,' 1882-84, p. 145 C.

⁵ ' Report of Progress Geol. Surv. Can.,' 1882-84, p. 140 C.

and at a distance so great from the nearest point of origin, is to me inexplicable except on the hypothesis that these erratics were brought here at one time by a single ice-raft or iceberg. Further to the north, and also not far from the base of the mountains, Doctor (now Sir James) Hector has described a line of very large red granite boulders at a height of 3,700 feet which it is probable may be attributed to the same time.[1]

The conditions implied by the existence of this latest drift of heavy boulders, while differing from those under which the boulder-clay was laid down, appear to require, as in the case of the boulder-clay, a continuous sheet of water extending from the westward front of the great Laurentide glacier, wherever that may have been situated at the time, to the positions of the erratics found; or, in other words, the transport of material from the east or north-east is not confined to the period of formation of the boulder-clay, but was continued and became even more pronounced at a later stage. It is impossible to satisfactorily explain the distribution of these erratics by systems of local post-glacial lakes. Dr. A. Geikie refers to the occurrence of similar large scattered boulders in Great Britain as indicative of floating ice,[2] and evidence of this kind is to be found everywhere on the Great Plains.

In relation to the particular point here in review, is the occurrence in the glacial drift of the plains of numerous erratics and much gravelly material referable to the Silurian and Devonian limestones which outcrop along the western base of the Laurentian axis or plateau, but are not anywhere met with over the area of the plains further west. These limestone outcrops now occupy a relatively very low position (say between 700 and 800 feet), but the abundance of their debris found far to the west tends to confirm a belief in the low level of the western part of the plains during the greater part of the Glacial period, and possibly to strengthen the supposition that the Laurentian plateau, with its flanking regions, was during that period correlatively elevated. The great bulk of the drift from these limestones is, however, confined to the second prairie steppe at heights less than 2,000 feet, while it runs out rapidly at higher levels, and is found in small quantity only above 3,000 feet.[3] This circumstance seems to show that the distribution of the drift material was largely a matter of level, and is thus again more easily explicable by water transport than by that of a great ice-sheet, by which it would appear to be reasonable to suppose that included material might be carried in equal proportion to its terminus. The present height of the western edge of the greater part of the limestone drift, may thus eventually be found helpful in assisting in the determination of the respective levels, at the time of its deposition, of the Laurentian plateau and Great Plains.

In earlier publications on the glaciation of the Great Plains, I have alluded to the general admixture in the drift of the plains of Laurentian and other eastern debris with material from the Rocky Mountains on the west, as a further evidence of the existence of a great sheet of water on which ice derived from different sources was borne in different directions.[4] The subsequent discovery by Messrs. McConnell and Tyrrell of local, and now isolated, patches of conglomerate, capping the Cypress and Hand-Hill Plateaux, composed of Rocky Mountain material and referable to the Miocene period, has since

---
[1] "Exploration of British North America," p. 224.   [2] "Text Book of Geology," 2nd ed., p. 897.
[3] "Geology and Resources of the Forty-ninth Parallel," pp. 246-247.
[4] "Geology and Resources of the Forty-ninth Parallel," p. 257. 'Quart. Journ. Geol. Soc.' vol. xxxi, p. 621. Compare also this paper, p. 14.

complicated this question, and has to some extent weakened the force of the evidence referred to, by shewing that much of this Rocky Mountain material (which I had designated in a general way as *Quartzite Drift*) has probably been secondarily derived from the Miocene conglomerates. The earliest opportunity was embraced of alluding to this new factor in the problem,[1] which, however, does not appear to me to be such as to invalidate the general argument derived from the mixture and overlapping of Laurentian and Rocky Mountain drift on the plains. The places above named are the only ones in which the Miocene conglomerates have been discovered, and their coarse character is such as to lead to the belief that they were formed in Miocene river-valleys or in the estuaries of rivers, and that the areas covered by them were never large. The existence of such conglomerate-covered areas appears in fact to have been sufficient to determine the existence of the plateaux mentioned and the limited size of these is probably connected with the actual small dimensions of the original deposit.

The Cypress and Hand-Hills Plateaux are situated at distances of 230 and 120 miles respectively from the base of the Rocky Mountains, while the material found in the glacial deposits and derived from the same mountains, is recognizable in these deposits, near the forty-ninth parallel, eastward to the 101st. meridian, 580 miles from its source and only 200 miles from the nearest part of the base of the Laurentian plateau.[2] It may be added, that quartzite from the Rocky Mountains may occasionally be noted in the glacial deposits in the form of boulders of considerable size, such as are not found in the Miocene conglomerates, while, moreover, a similar mixture of Laurentian and Rocky Mountain drift occurs much further north in the Athabasca and Peace River basins, where no trace whatever of Miocene beds of any kind has yet been observed; the Rocky Mountain material being there likewise found at times as boulders of some size.[3]

While therefore I am not prepared to insist on the absolute character of this evidence, it would appear that no discoveries which have yet been made, are such as to negative its general accuracy, the limited and local character of the Miocene conglomerates bearing but a minute quantivalent proportion to the wide spread of the Quartzite Drift.

Because of the absence in the area of the Great Plains of rocks capable of affording direct proof by their striation of the direction and character of the movement of the glaciating agent, I may be pardoned, before leaving this subject, for alluding to yet a further instance derived from the transport of material. The rocks forming the central part of the Three Buttes or Sweet-Grass Hills (which have already been mentioned), are of volcanic origin and differ from all others met with in the glaciated area of the Great Plains. A detailed study of the distribution of fragments of these rocks might therefore possess considerable importance in connection with the direction and manner of transport of material generally. We are as yet able to refer only to incomplete observations on this point, but even these appear to have some importance. In the immediate vicinity of the Buttes, trappean rocks derived from their summits are freely mingled with erratics of eastern and western origin, but at some distance to the east, north and south such fragments are rarely found. A few specimens of these rocks were, however, discovered at a

---

[1] 'Report of Progress Geol. Surv. Can.' 1882-84, pp. 142 C, 143 C.
[2] "Geology and Resources of the Forty-ninth Parallel," p. 225. 'Quart. Journ. Geol. Soc.,' vol. xxxi, p. 613.
[3] 'Report of Progress Geol. Surv. Can.,' 1879-80, p. 140 B. 'Quart. Journ. Geol. Soc.,' vol. xxxvii, p. 277.

point sixty miles due east. Fragments are also moderately abundant in the drift seventeen miles north of the East Butte and seven miles or more to the north of the West Butte.[1] The country to the south and south-west of the Buttes, being in the northern part of Montana, has not been examined by me, but it would appear to be in this direction, that the main stream of debris from the Buttes must be looked for.

The unoxidised and unweathered character of the material of the boulder-clays of the plains, may be quoted as a further evidence in favor of their sub-aqueous origin; and wherever they have not been manifestly subjected to recent subaërial action, this character is always apparent. They frequently contain great numbers of more or less completely rounded shaly fragments or pebbles, derived more particularly from the clay-shales of the Pierre group of the Cretaceous, which so soon as they are subjected to the weather, split up and crumble to pieces. Boulder-shaped masses consisting of sand have also been found in railway cuttings near Medicine Hat, on the South Saskatchewan, which it is difficult to account for except on the supposition that such sand, in a wet state, had been frozen into solid masses, which sank to the bottom of water and were immediately enclosed by earthy or clayey material there.

The direct evidence of submergence of the plains afforded by terraces and analogous phenomena may next be briefly alluded to. Terraces found along the river-valleys of the plains, with those met with in the valleys of the Rocky Mountains, may here be omitted from consideration, as being for the most part due to local circumstances, such as post-glacial river-erosion and the stoppage of mountain-valleys by local-glaciers. Terraces evidently free from suspicion in these respects are, however, found in considerable number and often well developed, in the higher part of the plains less than one hundred miles from the base of the mountains, and among the foot-hills. On this point I may quote as follows from my "Report on the Region in the vicinity of the Bow and Belly Rivers." "On approaching the mountains, however, true terraces of a more significant character present themselves in many places. Terraces in the entrance to the South Kootanie Pass at a height of about 4,400 feet have already been described in my Boundary Commission Report. In the valleys of Mill and Pincher Creeks, and those of the Forks of the Old Man, east of the actual base of the mountains, wide terraces and terrace-flats are found, stretching out from the ridges of the foot-hills, and running up the valleys of the various streams. Actual gravelly beaches occasionally mark the junction of the terraces with the bounding slopes, and they have no connection with the present streams, which cut through them. Their level varies in different localities, but the highest observed as well characterized, attains an elevation of about 4,200 feet. In the Bow valley near Morley (4,000) and thence to the foot of the mountains, similar terraces are found which are quite independent of the modern river; and in the wide mouth of the Kananaskis Pass a series of terraces was seen from a distance, which must rise to an elevation of at least 4,500 feet."[2]

Many years ago Sir James Hector, wrote as follows of the higher western border region of the Great Plains:—"On approaching the Rocky Mountains, the extreme regu-

---

[1] "Geology and Resources of the Forty-ninth Parallel," pp. 240, 241. Also MS. notes of 1883.
[2] 'Report of Progress Geol. Surv. Can.,' 1882-84, p. 146 C; see also "Geology and Resources of the Forty-ninth Parallel," p. 244; 'Quart. Journ. Geol. Soc.,' vol. xxxi, p. 618.

larity with which these deposits have been terraced by the retiring waters, at once attracts attention. * * * The terraces may be considered as ranging on the east side of the Rocky Mountains from 3,500 to 4,500 feet above the sea."[1]

To the east of this now more elevated part of the plains, evidences of water action of a similar character are to be sought rather in the existence of extensive tracts of almost perfectly level prairie, which occur at various elevations, and which may be regarded as more or less perfect examples of " plains of marine denudation " impressed on the soft and originally not very unequal surface of deposition. From this category, level plains formed by the infilling of post-glacial lakes, are of course excluded. Some evidences of water action are, however, also to be found here along the flanks of projecting plateaux. One distinct terrace of this kind is noted by Mr. McConnell on the Cypress-Hills Plateau, at a height of 3,700 feet,[2] while the south-eastern front of the Rocky-Spring Plateau, a few miles south of the international boundary, is strewn with Laurentian and Huronian erratics in a manner highly suggestive of an ancient shore line,[3] and the upper limit of the drift on the northern slope of the plateau, at the height noted on a former page, shows shingly materials resembling old beach deposits.

In order to explain the covering of the Great Plains with boulder-clay and other glacial deposits in conformity with the theory of the deposit of these from a great ice-sheet, we must admit that a great confluent glacier spread continuously across these plains from a gathering-ground situated on or beyond the corresponding portion of the Laurentian plateau. It has been suggested by some authors that such a glacier extended across the plains only as far as the line of the Missouri Côteau, and that a great lake, held in on one side by the front of the glacier, stretched thence to the western margin of the drift, a further distance of 300 miles. Such an explanation, however, appears to me to be inadmissible on account of the practical identity of the phenomena and deposits met with to the east and west of the Côteau, and more particularly because of the homogeneous character of the boulder-clay deposit of the two areas.[4] Others[5] have advanced the consistent, though bolder view, that the supposed great confluent ier reached quite to the western limit of the drift, or at least as far as the margin of the ognized boulder-clay, which does not fall far short of this limit. Holding this view, Mr. Tyrrell, in the light of facts still requiring the action of water in that part of the plains near the base of the Rocky Mountains, supposes that here at least lakes may have occurred in front of a great ice-sheet.[6]

On either of these forms of the hypothesis, we must admit that the plains were extensively covered by water at the inception of the Glacial period, (when the Saskat-

---

[1] " Exploration of British North America," p. 222.
[2] ' Annual Report Geol. Surv. Can.,' 1885, p. 74 C.
[3] ' Report of Progress Geol. Surv. Can.,' 1882-84, p. 148 C.
[4] There is no such general difference as between the drift deposits to the east and west of the Missouri Côteau, in this region, as to admit of the separation of the latter from the former and their inclusion as a whole under the designation of attenuated border deposits (Chamberlin) or fringing deposits (Lewis and Wright.) My earlier examinations of the plains near the forty-ninth parallel, being, to the west of the Missouri Côteau, principally confined the higher par'ts of the region, appeared to afford some indication that a general difference might be established, but subsequent more extended observations have not borne this out.
[5] Particularly Messrs. Upham, Wright and Tyrrell in publications already referred to.
[6] ' Bull., Geol. Soc. Am.,' vol. I, p. 401.

chewan gravels, were laid down); again in the interglacial episode the deposits of which separate the two boulder-clays; yet again during the principal period of glaciation, (when the supposed glacier-dammed lakes were formed); and finally, we are called on to suppose that large parts of the surface were covered by post-glacial lakes, of which the latest and lowest in level was that which Mr. Upham has named Lake Agassiz.

It is thus to account for the boulder-clays alone that we are on any hypothesis asked to admit the passage of a great glacier over the plains, and in view of the circumstance that no positively definite distinction in regard to physical characteristics has as yet been established as between boulder-clays of supposed sub-glacier origin and similar deposits produced by ice-laden water,[1] I cannot but think that the assumption of such a glacier invasion as that supposed is scarcely warranted or required by the observed facts.

Admitting for the moment the existence of a glacier co-extensive with the boulder-clay deposits of the Great Plains, we may note the required dimensions and movement of such a glacier. The main direction implied by the abundant Laurentian and Huronian drift of all parts of the glaciated area may be stated as from north-east to south-west. While it is possible that the principal direction of transport was more directly from the east or north, it must have lain between these bearings, and its precise direction is not important in the present connection. Whether the supposed great glacier originated on the Laurentian plateau or crossed it from some further point, it must at least be assumed to have moved continuously from the south-western edge of the Laurentian area to the furthest point at which material attributed to its action is found; or in other words, each identical transverse section of the glacier, charged or covered with Laurentian debris, must have been bodily pushed forward at least to the western edge of the boulder-clay deposits. This involves the conception that the glacier was thrust forward for a distance of over 500 miles from the margin of the Laurentian as a minimum, for no glacier-mass produced upon the surface of the plains and merely reinforced by accessions of ice from the Laurentian plateau will account for the carriage of the material from the latter.

In earlier discussions of this subject, I have assumed that a great glacier-mass thus moving across the plains would have to ascend their present gradual upward slope to the south-westward,[2] but if I am correct in the belief now entertained that the actual inclination of the plains was produced at a time subsequent to the main period of glaciation, we have to suppose only that the glacier was thus pushed forward over a nearly level surface, and we may admit that it was possibly assisted in its motion by a greater contemporaneous elevation of the Laurentian plateau. Even under such conditions, however, the glacier must have met with serious impediments, not the least of which is the escarpment of the Porcupine, Duck, Riding and Pembina "Mountains,"[3] composed of the cut-off edges of the Cretaceous strata, which faces the Laurentian plateau at an average distance of 130 miles from its margin and rises to a height in places of more than 1,000 feet above the intervening low tract. It is true that Mr. Tyrrell, while accepting the existence of an almost universal glacier such as that above indicated, quotes observations in favour of the belief that the flow of the ice descending from that part of

---

[1] Cf. Wright, "Ice Age in North America," p. 116.
[2] "Geology and Resources of the Forty-ninth Parallel," p. 261; 'Quart. Journ. Geol. Soc.,' vol. xxxi, p. 620.
[3] 'Geological Magazine,' Dec. ii, vol. v. p. 211.

the Laurentian plateau which is more or less continuously screened from the plains by this escarpment, on reaching the intervening hollow turned to a southward or southeastward direction.[1] These observations of direction of striæ, however, render it only the more difficult to account for the flow of the supposed great *mer de glace* across the plains, for if the ice from this nearest and most obvious source was thus largely or altogether turned aside, for a width of about 300 miles measured along the base of the Laurentian feeding-ground, the deficiency thus produced must obviously have been made up by a greater accession of glacier-ice to the plains from the north, which ice, crossing the fifty-third parallel with a width of about 500 miles, succeeded in attaining the forty-ninth parallel with a width of 600 miles and still with a very great thickness.

The local contributions to such a glacier by precipitation over the area of the southern part of the Canadian plains would be insignificant, while the loss by melting must be supposed to have been great, as the region in question is in the central part of the continent, and must then as now have been one of small rain or snow-fall.[2] The hypothesis of such a vast mid-continental glacier and the effects attributed to it appear to be particularly opposed by this circumstance.

In addition to these more general aspects of the conditions implied by the theory of a *mer-de-glace* extending over the area of the Great Plains, reference may now be made to still another circumstance which to me appears in itself to be destructive of the hypothesis, this being dependent on the effect which such an ice-mass must have produced on materials over which it passed. Doubt has been felt and expressed by competent authorities as to the possibility of the formation to any considerable extent of till or boulder-clay beneath the mass of an ice-sheet in motion.[3] It is obvious on *a priori* grounds as well as on reference to known instances of glaciation on a large scale, that the principal effect of a moving ice-sheet or glacier is in the direction of the removal of all incoherent deposits and the baring of the underlying and more resistant rock-surface, the face of which is striated and smoothed, or even abraded and channelled, when the ice-covering has been heavy or its action long continued. It is thus only under somewhat exceptional circumstances and locally that till can be proved to have originated beneath a glacier-mass, or that antecedent deposits of a non-indurated character can be shown to have maintained their position below such a mass. Such instances have, however, been made the basis for an almost general theory of the origin of till or boulder-clay, as a bottom-moraine, and in some instances this has been carried so far as to practically relegate the whole of such deposits to this origin. The extension of such a theory to the vast area of the plains appears to subject it to a strain greater than it is capable of bearing, and such an extension can in fact be mentally approached only by a hypothetical series of increasing approximations, beginning with the minor observed instances which have been referred to and leaving other circumstances out of consideration.

Prof. James Geikie, in offering an explanation of the preservation of certain deposits under till on the hypothesis that the latter was itself formed beneath a great glacier, writes :—"In the open lowlands and in the broad valleys, where the ice-sheet would

---

[1] 'Bull., Geol. Soc. Am.,' vol. i, p. 401.

[2] Cf. W. J. McGee, "On Maximum Synchronous Glaciation." 'Proc. Am. Assoc. Adv. Sci.,' Boston, 1880, p. 447. Also Von Wadekoff, 'Proceedings Geol. Soc., Berlin,' 1881.

[3] Cf. 'American Journal Science,' III, vol. xxxiv, p. 52, where Mr. O. P. Hay attempts to explain such formation.

advance with diminished but more equable motion, we come upon widespread and often deep glacial deposits, and now and again with interglacial beds; while over regions where the gradually decreasing ice-sheet crawled slowly to its termination, we discover considerable accumulations of till, often resting upon apparently undisturbed beds of gravel, sand and clay.[1] It is not, however, alone with an ice-sheet in this moribund state that we are required to deal in the region here in question. The ice-sheet, if it was at any time co-extensive with the boulder-clay, must, as already stated, be supposed to have moved bodily across the Great Plains for a distance of over 500 miles. If it be admitted that only one length of 500 miles of exceedingly thick ice traversed the eastern part of the plains in its south-westward passage, we might expect to find some very remarkable traces of its progress impressed on the underlying soft Cretaceous rocks. As, however, the Glacial period was of considerable duration, it is necessary to suppose rather that the eastern part of the plains must have been subjected to the passage over them of at least several thousand miles of such ice, which was not in a decaying condition, but after leaving the Laurentian plateau was about to begin its long journey across the plains. The composition of the boulder-clay throughout the plains implies the incorporation of a large proportion of relatively local material into its mass, so that it cannot be argued that the eroding power of the ice was insignificant; yet no such continuous or extensive ploughing up and channelling out of the subjacent rocks as might be supposed to result from the passage of such an ice-sheet has been observed. On the hypothesis of the production of the glacial deposits of the plains by floating ice, the local material may be supposed to have been supplied by the occasional impact upon the bottom of the grounding bergs, by the disintegration of prominences which stood at times above the surface, and by the action of field ice upon these.

It is, however, in the western part of the plains that the gravest difficulties occur in connection with the theory of a vast ice-sheet moving over subjacent deposits. As noted in former pages, we find there the immediately pre-glacial and generally quite unconsolidated Saskatchewan gravels, with associated bedded sands and silts, spread pretty generally and in thin layers over considerable areas of a country of very low relief. These lie beneath the lower boulder-clay and are at a distance in places of nearly 200 miles from the western margin of the glacial deposits. The thickness of the ice-sheet which would, according to theory, be supposed to have been pushed westward for about 200 miles over these beds (estimating this from the height of the upper edge of the glacial deposits on the neighbouring Cypress Hills plateau) must have been from 1,500 to 2,000 feet! Above the lower boulder-clay are the sandy and silty interglacial beds with peat, which though not observed to be more than about thirty feet in thickness, and often much less, can be traced in nearly continuous sections for forty-five miles or further along the Belly River. Above these lies the upper boulder-clay, which, if it in turn be attributed to a second advance of a great ice-sheet, involves a belief that a thickness of ice similar to that above noted traversed these interglacial deposits, part of which lie about one hundred miles within the edge of the glacial drift. Though locally wanting, neither the pre-

---

[1] 'Geological Magazine,' Dec. ii. vol. v, p. 76.

glacial nor the interglacial deposits referred to have been observed to be disturbed or contorted.¹

From what has already been said with respect to the Cordilleran region, and more particularly in connection with the meaning which the White Silt formation appears to have in that region, it seems probable that the water by which the northern part of the Great Plains is supposed to have been flooded, was in connection with that of the sea.² In discussing the results of my earlier investigations of the superficial deposits of this part of the plains, in reference to a theory of their submergence, I have stated, that after a certain stage the waters entering from the north and south must have formed an open strait between the Arctic Ocean and the ocean to the south.³ This was written, however, under an assumed limitation implying an equal subsidence of the continent, and at the time no satisfactory information was available respecting the position of the margin of the glacial deposits in the corresponding western part of the United States, such as has since been supplied by the work of Chamberlin, Salisbury, Todd, Wright, McGee, Upham and others. The result of these new facts appears to show that instead of opening broadly southward as well as to the north, any body of water covering the northern part of the Great Plains could have had only a tortuous and comparatively narrow communication with the sea to the eastward, round the front of the great confluent Laurentide glacier, and that even this communication was probably formed only at the time during which the plains stood at the lowest level indicated by the greatest spread of the drift deposits. If such conditions may be assumed as probably representing the facts at the time, they go far toward explaining one of the greatest difficulties against the acceptance of the hypothesis that the waters by which the plains were flooded were in communication with those of the sea. The difficulty alluded to is the complete absence, so far as yet ascertained, of the remains of marine organisms from the glacial deposits. While prolonged weathering and the action of sub-aërial waters might result in the removal of calcareous organic remains from certain parts of these deposits, the condition of much of the boulder-clay, together with the occasional actual occurrence in it of fragments of Cretaceous or Laramie shells, is such as to show that any contemporaneous molluscs might have been preserved. If, however, the body of water in question, though communicating with the sea to the northward, was almost throughout closed to the south and in receipt of large quantities of fluvial water, it may well have been in great part brackish, if not almost entirely fresh. Adding to this the conception of its frigid temperature due to the great abundance of ice with which it must have been laden, and the vast amount of fine sediment which must have been carried into it by sub-glacier streams, it will be apparent that the conditions were singularly inimical to the existence of life of any kind, whether that characteristic of salt or fresh water. Somewhat similar conditions, though on a much

---

¹ Prof. J. W. Spencer has lately adduced much additional evidence to show that modern glaciers, near their terminations, produce little effect even on loose material beneath them, ('Trans. Roy. Soc. Can.,' vol. v, sec. iv, p. 89.) but the mode of action of a glacier at its decaying extremity and when in contact with materials in a position to acquire more or lose heat from the atmosphere, and by radiation and conduction from neighboring warmer bodies, must be very different from that which would be found far back beneath the mass of even a small glacier.

² It may still, however, be admitted as possible, that a great lake was in some manner produced, in the region of the plains, with a height somewhat exceeding that of the sea.

³ "Geology and Resources of the Forty-ninth Parallel," p. 255

smaller scale and without the adjunct of glacial waters or glaciers, are found to-day in the southern extremity of Hudson Bay, where, as Mr. A. P. Low informs me, marine life is almost entirely absent, the water being nearly fresh and clouded with mud derived from the large entering rivers and from the action of the waves upon the shoal, earthy shores.

In regard to the position which the front of the Laurentide glacier or confluent ice-cap of the Laurentian plateau may have occupied when at its maximum, I need here say only, that information lately obtained by Mr. Tyrrell,[1] as I read it, appears to show that this ice-front reached the escarpment of the Riding and Duck Mountains with a thickness of about 2,000 feet; while the occurrence in some places in the Red River valley of a peculiar indurated "hard-pan" as the basal member of the glacial deposits (as in the boring at Rosenfeldt) which has much the appearance of a sub-glacier deposit or true till, leads to the belief that the great glacier may at one time have occupied a considerable part or the whole of that valley to the north of the international boundary.

When the study of the superficial deposits of different parts of Europe and America was for the first time seriously begun it was endeavoured to explain the phenomena entirely by diluvial action, and when the evidence of ice-action became insuperable, ice-bergs and floating ice only were at first admitted as factors. Since that time the pendulum of opinion appears to have swung to an opposite extreme, and the energies of the majority of investigators have been expended in endeavouring to account for the varied facts of what has become definitely known as the Glacial period, almost exclusively by the action of great confluent glaciers. From this extreme point, the pendulum may now be supposed to have returned so far, as to leave the hypothetical North Polar Ice-cap almost without an advocate, but at what position it may eventually come to rest time and the further progress of research alone can decide. I am aware that some of those who have accepted what I may perhaps be pardoned for characterizing as extreme views as to glacier action, have more or less completely, and to their own satisfaction at least, solved all difficulties opposed to the action of glacier-ice, such as those presented by the facts met with over the Great Plains, by the application to these of their single universal menstruum. For myself I need only say that I have endeavoured to approach the subject of the glaciation of the north-western part of the continent, here reviewed, untrammeled by *a priori* theories, and with some personal familiarity in the field with nearly all parts of the region dealt with.

Little reference has been made to the mass of literature which exists on the glacial phenomena of the eastern part of the continent, as the western region here particularly referred to, appears to be sufficiently vast to be considered on its own merits, while the conditions met with in it are in many respects different from those found in the east.

---

[1] 'Annual Report Geol. Surv. Can.,' 1887-88, p. 13 E; 'Bull. Geol. Soc. Am.,' vol. i, p. 399.

EXPLANATION OF MAPS Nos. 1 TO 4.

No. 1.

Sketch-Map, shewing approximately the extent of the main area of the Earlier Cretaceous sea.—Queen-Charlotte-Islands and Kootanie Period.

No. 2.

Sketch-Map, shewing approximately the extent of the main area of the Cretaceous sea during the Dakota Period. The regions unsubmerged in the Cordillera were probably much larger than in the Earlier Cretaceous, but few of them can as yet be defined. In the Benton and Pierre Periods the Cordilleran region formed a more nearly complete land-barrier between the submerged areas of the Great Plains and the Pacific, and in the intervening Belly-River Period, the condition of the part of the Great Plains included in the map, appears to have approached that found in the Laramie.

No. 3.

Sketch-Map, shewing approximately the principal submerged areas during the Laramie Period.

No. 4.

Sketch-Map, indicating in a general way the portion of the Cordilleran Region, which is supposed to have been more or less continuously ice-covered during the first maximum of glaciation. The principal directions of flow of the Cordilleran Glacier, are indicated by the curved blue lines. The red line shews the maximum westward spread (at a later epoch) of material derived from the Laurentide Glacier.

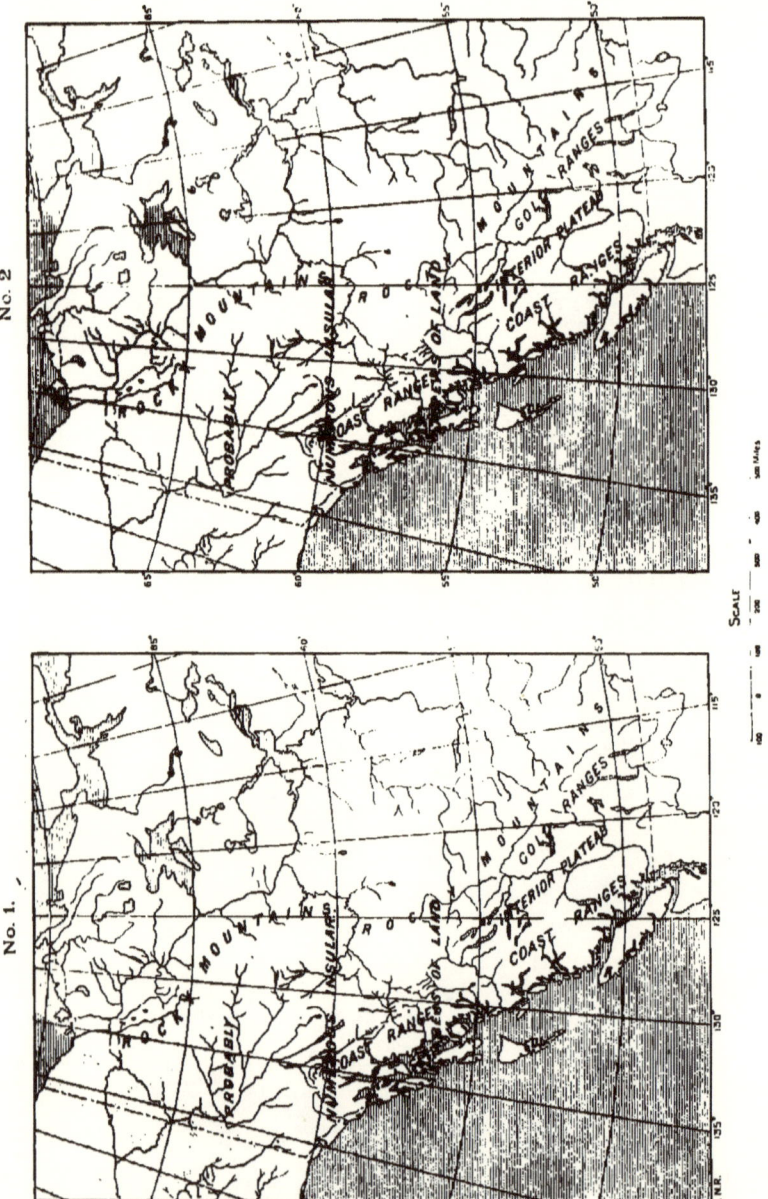

To illustrate Dr. G. M. Dawson's Paper on the Rocky Mountain Region of Canada.

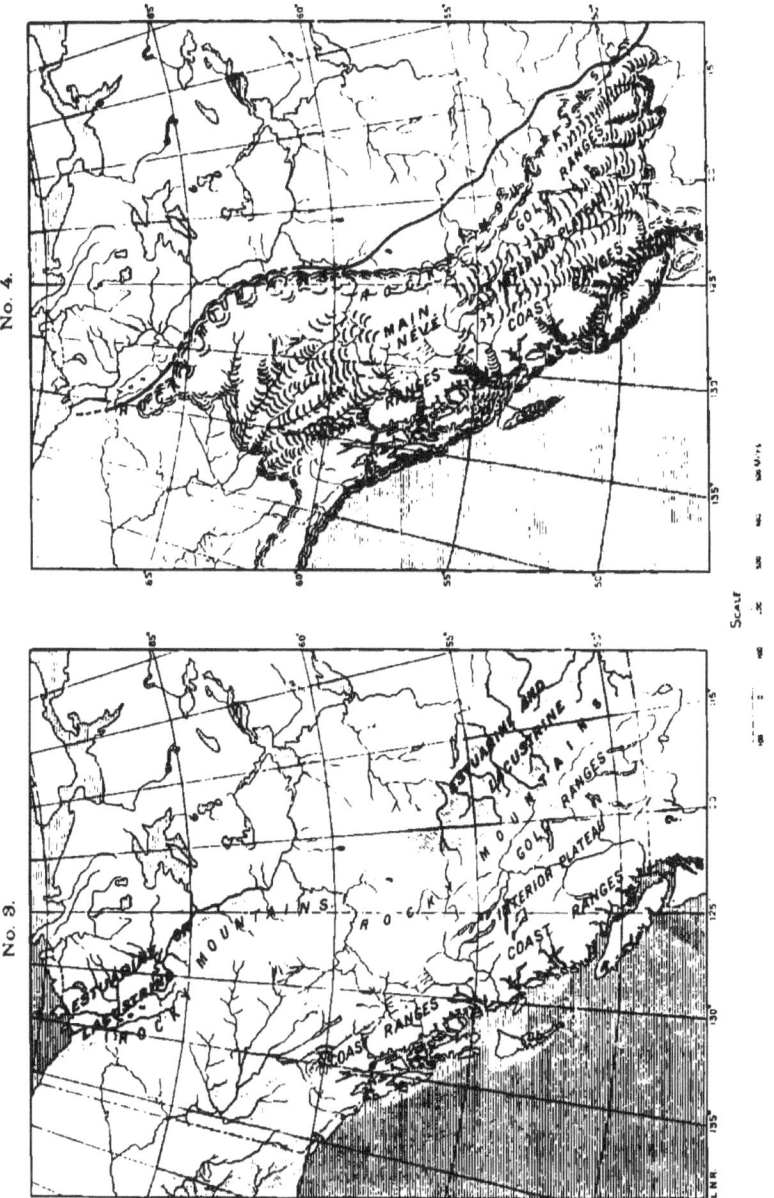

To illustrate Dr. G. M. Dawson's Paper on the Rocky Mountain Region of Canada.

To illustrate Dr. G. M. Dawson's Paper on the Rocky Mountain Region.

www.ingramcontent.com/pod-product-compliance
Lightning Source LLC
Chambersburg PA
CBHW020325090426
42735CB00009B/1415